WITHDRAWN

PRESENTED TO
CODY MEMORIAL LIBRARY
BY
KELLOGG FOUNDATION

HUMAN GROWTH

HUMAN GROWTH

The Story of How Life Begins and Goes On

Based on the Educational Film of the Same Title

By LESTER F. BECK, Ph.D.

ASSOCIATE PROFESSOR OF PSYCHOLOGY,
UNIVERSITY OF OREGON

with the assistance of Margie Robinson, M.A.

75758

HARCOURT, BRACE & WORLD, INC. • NEW YORK

Cody Memorial Library
Southwestern University
Georgetown, Texas

COPYRIGHT, 1949, BY LESTER F. BECK

All rights reserved, including the right to reproduce this book or portions thereof in any form.

PRINTED IN THE UNITED STATES OF AMERICA

C
612
B388h

ACKNOWLEDGMENTS

THIS BOOK and the film, *Human Growth,* are the result of extensive psychological research which has been sponsored over a period of years by the E. C. Brown Trust, a foundation established in Oregon a decade ago for social hygiene education.

Appreciation and gratitude are expressed here to those thousands of mothers and fathers who participated as critics in the early testing of the film, *Human Growth,* and to the several hundred more who, with the help of their teen-age sons and daughters, evaluated a pre-publication edition of this book.

Appreciation is also expressed to Eddie Albert and John E. Fletcher of Eddie Albert Productions for their contributions to both the film and the book. The illustrations are the work of John Hubley and Phil Eastman of United Productions of America, and are reproduced through the courtesy of the E. C. Brown Trust.

Copies of the film may be borrowed or rented from many sources in the United States and Canada, including state universities, state health departments, educational film libraries, and local school systems. Full information about means of obtaining the film, may be had by writing to the E. C. Brown Trust, Education Center Building, 220 S. W. Alder Street, Portland 4, Oregon.

CONTENTS

HUMAN GROWTH

INTRODUCTION

THIS BOOK is written for boys and girls just entering their teens who wish to know about the changes that take place in their bodies as they grow up. It is also written for those adults who have the important task of helping young people achieve a wholesome view of life and growth. This book, *Human Growth,* covers the same material as the widely known film, *Human Growth,* but the book is able to go into considerably more detail and should be read for itself as well as in connection with the film. It tells in simple language how growth begins and how it continues, and it gives a full explanation of the physical changes during adolescence and their meaning for adult life.

Each chapter ends with a set of questions and answers. These are actual questions asked by children after they saw the film and read the book. The purpose of the questions is to give the young reader an idea of the many facts about growth and reproduction other boys and girls want to know. They also are meant to provide a basis for frank and intelligent discussions with adults about how life begins and how it continues.

The book—like the film—stresses normal, healthy development of the body. No special effort is made to describe the social and emotional changes or the moral and ethical problems that face the child in the process of growing up. These topics, it is felt, are more closely related to motives—to internal forces that cause human beings to act in certain ways—than to physical growth. The main object of the book is to give young people a clear understanding of birth and growth without at the same time involving them with questions of right and wrong. These questions should properly be discussed in the family, the school and with the people who have charge of religious instruction.

This book does not claim to be the whole answer to the problem of sex education. It is concerned with only a small part of that problem. The child's attitude toward its home, its parents, its brothers and sisters, members of the opposite sex—all these are important to sex education. Some of these attitudes are shaped by the family, others by the church, still others by the school and related social groups.

The film *Human Growth* and this book came into existence in the following manner. In 1939, the late Dr. E. C. Brown gave to the University of Oregon a sum of money for research and experimentation in new ways of social hygiene education. Nine years later the film *Human Growth,* one of the projects made possible by this grant, was completed and ready for general distribution. This book was written shortly afterward, and has taken advantage of many of the techniques and methods which made the film so overwhelmingly popular. Book and film, therefore, are a joint project, both stemming from the same extensive, thorough research and planning.

Both the book and the film *Human Growth* have been shown to parents, teachers and professional workers for comment and criticism. The result was more than satisfying. Of the thousands of parents and teachers who have reviewed the film and the book, ninety-seven per cent have approved them for home and school use.

The film is as popular with grownups as it is with young people and is being shown throughout the United States and the world. The story of just how it was filmed is quite interesting. The actors were junior-high-school students who gave up part of their summer vacations to make the movie. A large share of the credit for the film's success should go to them. *Human Growth,* moreover, was produced in a regular school classroom instead of on a motion-picture set. All the equipment for lighting, photography and sound was set up inside the school building.

Unlike most movies, *Human Growth* was made to be shown in school, not at the local theater. It was planned this way to enable pupils in the classroom to talk over the material in the picture, and by so doing to learn about themselves and about the process of growing up. The movie tells the story of human reproduction and of the changes that take place as babies grow into children and children into young men and women. In classes where the film is shown, time is always left for a discussion. The teacher takes part in these talks, answering questions and supplying additional information.

This book covers the same facts of human growth as the film. Like the film, it answers important questions which young people ask about birth and growth; it also helps to explain what it means to be a father or a mother. Again like the movie, the book can be used for discussions between children and their parents and teachers. Thus the book, *Human Growth,* and the film, *Human Growth,* cover the same ground in almost the same way—their main difference is that one is a film and the other a book. Both contribute to mental health by a simple, honest telling of the story of human growth and by providing a sound basis for discussion with others.

CHAPTER ONE

GROWING...
GROWING...GROWN!

How a person grows after he is born and how the various glands in the body affect growth and other body functions

THE story of how human beings grow up is as dramatic as any tale ever written. The life plan which, in twenty-odd years, changes a tiny baby into a full-grown man or woman is one of the great wonders of nature. And it is taking place every day—every minute, in fact—among boys and girls throughout the world.

This process of growth has already started when the baby is born. But he has far to go, and much to learn, before he becomes an adult.

Very soon after birth the baby announces his presence to the world with a yell. He continues to cry, in fact, for quite a while. Since he cannot talk, a loud yell is his only means of letting people know what he needs. He wails to show that he's hungry . . . tired . . . cold . . . or uncomfortable.

Some young people are surprised and amused by the looks of a newborn infant. The baby is likely to be red

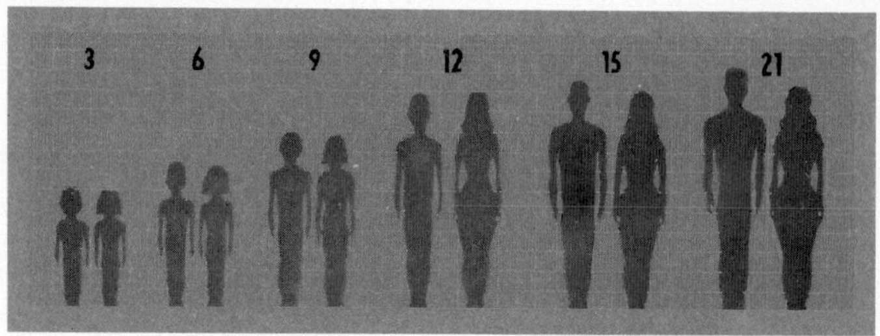

This chart shows growth changes in height and shape of the body for boys and girls from the age of three to twenty-one. Notice the marked change that occurs between the ages of nine and twelve.

and small, with a button of a nose and squinting eyes. He may or may not have hair slicked close to his head. With his wrinkled and screwed-up face, he looks something like a little old man. Before long, however, he begins to acquire a personality of his own.

Now let us follow an average baby through the first few years of life and see what he learns as he grows. Remember that some babies will do things sooner—some later—than others. The infant we are about to describe is just a sample. Actually, no one baby develops in exactly the same way as another.

Besides being very weak and helpless, the newborn baby can't see clearly for a month or more. When he is about four weeks old, he stares at people's faces. At two months he watches a toy swung in front of him and even smiles. His "speech" at this time consists mostly of gurgling noises.

At the age of four months, the baby has learned to shake his rattle. A month later he can pick up a spoon—though he still cannot use it to feed himself.

If a mirror is put in front of the baby at six months, he will touch the reflection as though he were reaching out to pat a living twin. By now the baby is plump, bright-eyed, and alert.

During these months he has been gaining in size and strength. There are two periods in the lives of human beings when growth is the most rapid. The first of these periods is immediately after birth. The second starts during the early teens.

By the time he is seven months of age, the fast-growing baby has developed sufficient bone and muscle to enable him to sit up alone. And he can play with toys—blocks, teddy bears, even his own toes. A month later he can stand up, tottering, with the aid of an adult. Soon, grabbing chair or table legs, he is clambering about the room.

Now comes the first birthday. This day is a milestone in the life of the baby—as well as in the lives of his parents. By now their child can very likely play pat-a-cake, wave good-by, stuff a cracker into his mouth. And he can probably say "ma-ma," "da-da," and other equally simple "words." Fascinated by his own voice, the baby keeps repeating the same sounds over and over again. He is practicing the sounds he will use later in everyday speech.

At eighteen months, most babies can pronounce

about seven words. From then on they learn so fast that by the age of two they can say from 250 to 300 words. By the time a baby is two and a half, most of these words can be understood. He is outgrowing the mixture of sounds and squeals which, for a while, made a baby language of his own. The more real words he learns, the more fun parents have with him; and it is not long before the grownups are listening to him recite nursery rhymes.

One of the proudest moments for the family comes when the baby takes his first wobbly steps. This performance usually starts at the age of fourteen or fifteen months. In his eagerness to reach some object or person, the baby often falls down. His legs are so short, however, that such falls do not do him any harm.

By his second birthday, the child can not only walk—he takes great delight in running all over the house. For several months he has known how to use the toilet. Now, with so many words at his command, he can tell his mother when he needs to go. At playtime, two-year-olds enjoy dolls and often carry them about the house.

At the three-year mark, the child has learned to feed himself, to help dress himself and to put on his shoes. At this age, boys and girls are also playing with other children and learning that life involves getting along with other people.

At four, children begin to ask endless questions. They can wash and dry their own hands and faces. They can also help their parents by running errands. Their speech has reached the stage of complete sentences. At

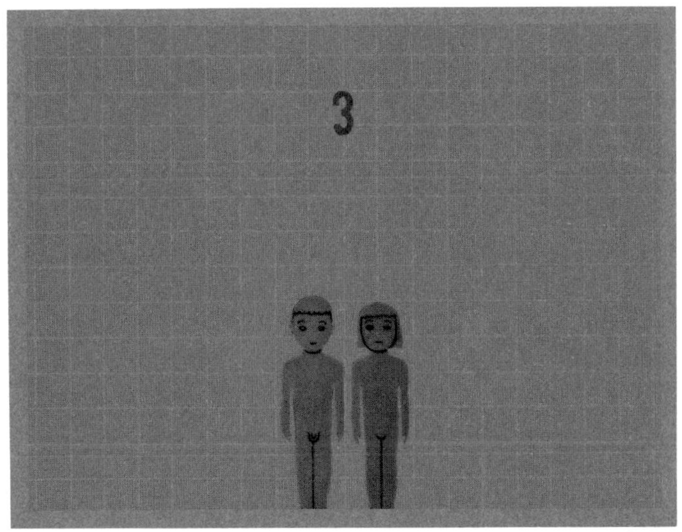

At the age of three, the head is round and the body short. The boy is a little taller than the girl.

five, they can not only dress without assistance, they can also pronounce between 1500 and 3000 words.

Activities such as running, climbing, throwing and catching a ball, jumping rope, skating, and skipping have developed by the time children first start going to school.

Don't forget that we have been talking about the development of an *average* baby. The times at which walking, talking, and other achievements start vary greatly from child to child. Growth does not follow a strict schedule.

One way of learning about human growth is to compare the size of boys and girls at different ages. At birth,

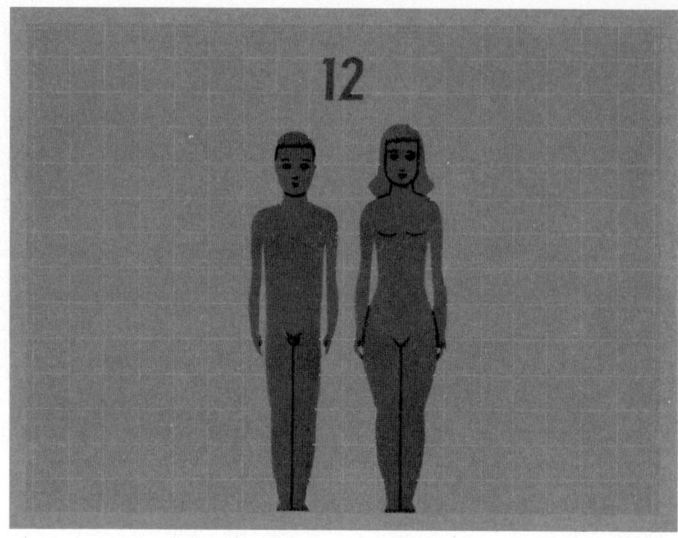

At twelve, the girl is larger than the boy and she looks more mature. This is because the girl grows faster than the boy between nine and twelve.

the boy and girl are more or less the same size, although the boy is apt to be slightly taller and heavier. During the first three years of life, both boys and girls gain very fast in height and weight. Thus three-year-olds are twice as tall as newborn infants and three times as heavy.

In childhood, between the ages of three and ten, the speed of growth slows down. While girls, on the average, grow a fraction less than two inches a year, boys grow slightly more than two inches. Thus most boys are still taller than most of their girl playmates. But here's a point to keep in mind: Not *every* small boy is taller than *every*

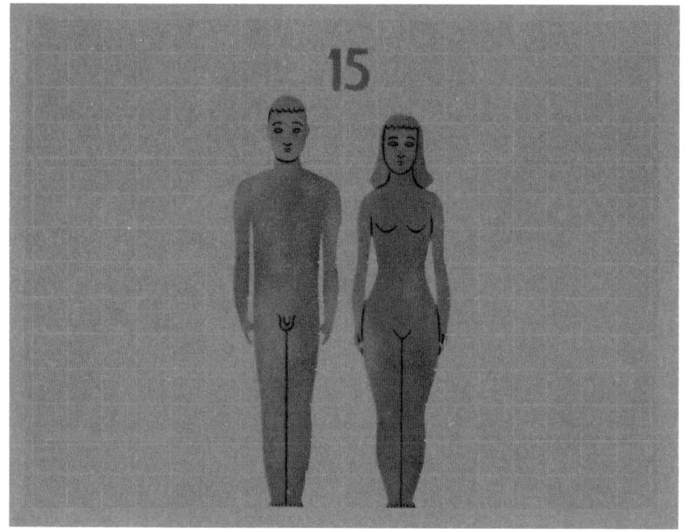

From twelve on, the boy usually grows faster than the girl so that by the age of fifteen the boy is again taller and heavier than the girl.

small girl. Food, sleep, fresh air, sunshine, exercise, size of parents and grandparents—all these things affect growth.

When they are ten or eleven years old, girls start growing more rapidly, passing boys in both height and weight. This "growth spurt" lasts until they reach thirteen or fourteen. Thus girls become physically mature about two years sooner than boys. This means that most girls of twelve and thirteen are bigger than most boys of the same age. Then it is the boys' turn. At twelve to thirteen, they start outgrowing their clothes at a surprising speed. By fifteen, they are equal to girls in size.

This is the period known as adolescence. In girls adolescence usually begins at the age of about twelve to thirteen, in boys at about fourteen to fifteen, and continues until they have reached maturity. Though the growth rate now slows down for girls, boys keep growing rapidly for another year or two. The time just described is the second period of rapid growth which occurs in all people. The first period, remember, came immediately after birth.

No two children, however, grow at precisely the same speed. Just as all human beings differ in appearance and behavior, so they differ in the way they grow. Take

This group of boys and girls are thirteen years old. They differ considerably from one another in physical maturity and size. Such differences are normal.

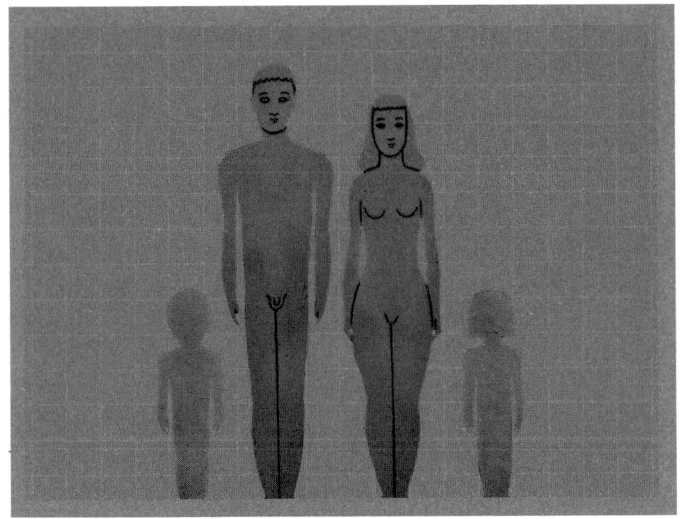

The adult, compared with the child, has a narrower face and head, and longer limbs and waist. The boy's shoulders are broader and his body more muscular, while the girl's body is more curved and feminine.

three fifteen-year-old boys, for example. One may still look like a child, with his straight slim build and small-boy features. Another may be in the gawky stage. Yet another may look like a grown man, even though he is just starting high school.

Growth also produces great differences in build. While one young girl is short and slender, another may be short and plump. Again, some boys and girls are tall and slender, others tall and heavy. Similar differences in shape and size can be seen in any group of adults.

At about twenty years of age, the average boy is five inches taller and fifteen pounds heavier than the average girl. By this time—the late teens or early twenties—physical maturity has been reached. At eighteen, girls are fully grown, while boys continue growing a little taller up to the age of twenty or so.

People often talk of growing *up,* as if adding inches to their height were the only change that occurred between infancy and adulthood. Actually, growth is a far more complicated business than simply growing *up*. The form, or shape, of the body changes, too—and in several ways. The head, the arms and legs, and the other parts of the body grow at different rates.

Take the head, for example. The cranium or bony part grows most rapidly before birth, while the child is still in its mother's womb. After birth the head continues to increase in size, until by the age of nine or ten it has almost reached its full growth. At birth the baby's head is so large—one-fourth as long as his whole body—that it's apt to make him look top-heavy. As an adult, however, the head is only about one-eighth of his total height.

The bones of the arms and legs are very short at birth and remain relatively short during childhood. Not until the age of nine or ten do they start to lengthen; then they continue growing during adolescence. They grow at such a pace, in fact, that a teen-ager's arms and legs may even appear out of proportion to the rest of his body, but this awkward stage does not last long.

The trunk—that part of the body between the neck and legs—develops in still another way. Long at birth, it grows fast for a brief time; then slows up during childhood and the early teens. The trunk lengthens to its full size only as the boy or girl reaches maturity.

A child has almost no waistline. A teen-ager, however, has a high waist, which makes him appear long-legged. Finally, as the trunk continues its growth, the young man or girl acquires a balanced, adult appearance.

Even the face develops in a special way. The features grow slowly in childhood, then more rapidly in adolescence. The upper part of the face—forehead, eyes, and nose—develops faster than the lower part. The chin and jaw are the last features to reach adult size. A four-year-old has a snub nose, and flatter lips and a smaller chin than he will have when fully grown. The adult, on the other hand, has a longer and more shapely nose, with the dip under the brow filled out, a higher forehead, a fuller mouth, and a firmer chin. These features, of course, are very rarely the same for any two people, unless they happen to be the kind of twins you can't tell apart.

Differences in the shapes of boys and girls become noticeable as soon as they enter their teens. Boys develop wider shoulders, longer arms and larger hands than girls. They are also stronger and use their muscles more. In various physical activities, boys as a rule become more skilled than girls.

As the girl approaches womanhood, her breasts be-

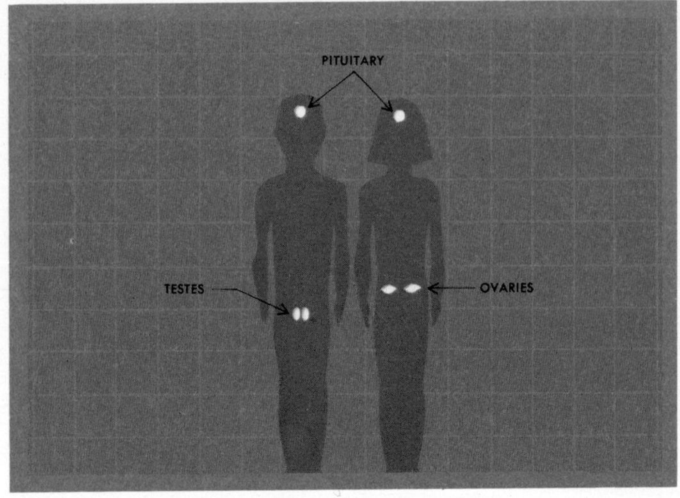

Growth is controlled by tiny organs within our bodies called glands. One of the most important is the pituitary gland, located in the head. Others are the testes and ovaries in the pubic region of the body.

gin to fill out, her hips grow rounder and wider. This feminine appearance accompanies certain changes in the glands of the body. These glands give boys a more rugged and masculine appearance.

Glands are tiny organs in the body, and they are extremely important. While some glands control growth and development, others regulate functions of the body. Now let us investigate the two main families of glands and the special tasks which they perform.

One group of glands, known as the duct glands, includes the sweat glands and the oil glands, located in the skin. As a person enters adolescence, these glands become more active. Hot weather, violent exercise, or excitement usually causes sweating. This perspiration sometimes produces an odor which can be prevented by daily baths and the use of deodorants.

Increased activity of the oil glands, combined with irregular eating habits, during this period of rapid growth, may result in skin troubles. Such disturbances vary from an occasional pimple on the face to a severe rash known as acne. Cleanliness, a well-balanced diet, sufficient sleep and exercise usually clear up this condition. In some cases, a doctor's assistance may be needed.

Another group of glands, known as the endocrine glands, manufactures within the body chemical substances called hormones. These glands—which answer signals from the nerves in various parts of the body—send their own hormones into the blood. The hormones may then act upon other glands. Though they are produced in extremely small quantities, hormones are very important to the process of growth. They act upon parts of the body in special ways, depending upon the glands from which they come. This message system—from the nerves to the glands, which in turn produce hormones—is most remarkable.

There are several different glands in the endocrine group. Each of them has a certain job to perform, and each one sends out hormones, the messengers which do the work.

75758

Cody Memorial Library
Southwestern University
Georgetown, Texas

The hormones from one gland, for instance, help the body to digest food in the way that will develop strong teeth and bones and firm muscles. Hormones from another gland enable the body to burn up sugar and transform it into energy. Yet another hormone enables us, in sudden excitement or fright, to receive an extra supply of energy, which allows the muscles to act with greater speed. We have all seen how a cat, when frightened by a dog, arches its back and stiffens its body. The cat is then ready for a fight. Inside its body the cat's hormones are at work, helping it to meet this emergency.

A most important gland in the endocrine family, though no larger than a pea, is the pituitary. It is located in the brain. One kind of hormone secreted by this gland is necessary for growth. Its function is to see that the body of a healthy, well-fed child attains its full size. Another pituitary hormone affects the work of other glands, especially the sex glands. These sex glands are located below the extreme lower part of the abdomen.

Sex glands differ in men and women. A man's sex glands consist of a pair of organs known as the testes (or singly, a testis). A woman's are called ovaries.

Although babies and children possess these sex glands, they do not become active until the age of puberty. Puberty marks the beginning of adolescence and means that human beings can reproduce their own kind. Puberty usually starts in the early or middle teens, in girls two years earlier than in boys. We shall hear more about the sex glands and adolescence in the next chapter.

The material just covered in this chapter calls forth many questions from young people. Here are some of them. You may think of others that you wish to ask your parents or your teacher.

QUESTIONS AND ANSWERS

Could something go wrong with a child so that he would stay a child's size all his life?

This is most unlikely. Almost all children grow normally and reach adult size. When something goes wrong with the pituitary gland, normal growth has been known to stop. In these very rare cases the person becomes a dwarf. When, on the other hand, the pituitary gland is active too long, the person can become a giant. However, these things happen very seldom.

If babies and small children don't smell of perspiration, why do those who are older?

Just before and during the early teens, certain glands in the body become more active. This is part of the preparation for adulthood. Among these glands are the sweat glands and the oil glands. They begin producing more oil and more sweat (perspiration) than during childhood. Large secretions of these substances may cause an odor, which can be prevented by frequent bathing and the use of deodorants.

Why do people have hair on their bodies?

We do not know exactly why hair grows on some parts of the body. The appearance of body hair, especially under the arms and in the pubic region, is one way of telling that people have reached sexual maturity, in other words, that their sex cells should be capable of reproduction.

Why do so many junior-high-school boys and girls have pimples on their faces? Can anything be done about it?

During adolescence the oil glands in the skin become very active. They produce a great deal of oil, which should leave the skin through tiny openings called pores (or, in regions where hair grows, through follicles). Because different parts of the body develop at different speeds, the pores cannot always keep pace with the work of these glands. The oil then collects in small pimples under the skin, which may swell and become irritated.

Frequent use of soap and water helps to get rid of the excess oil. Other aids may include a special diet, regulation of sleep and exercise, and—in severe cases—a doctor's care.

Why are some thirteen-year-old girls taller than boys of the same age?

As a rule, girls reach physical maturity sooner than boys. Girls start growing rapidly at ten or eleven, usually reaching their full height between eighteen and twenty. Boys, on the other hand, begin this growth spurt two years later than girls, and continue growing until they are twenty-one or -two. Thus thirteen-year-old girls have a "head start" on boys of the same age.

Why does a woman develop such a different shape from a man?

One of the functions of a woman's body is to bear children. Her hips become broader and her abdomen longer than those of a man in order that her body may have room for a baby. Her breasts fill out so that when she has given birth, she may provide her baby with milk.

Men's bodies are able to perform heavy work, such as digging and building. For such activities they have broader shoulders and stronger muscles.

Why can't newborn babies see very well?

Babies *can* see, but their eyes require time to grow accustomed to the world outside their mothers' bodies. At first, babies can't focus their eyes as adults can, nor can they distinguish colors. They can tell light from dark, but they cannot see very far. Gradually the eyes of babies start to co-ordinate, and what they see begins to *mean* something to them.

CHAPTER TWO

ABOUT OURSELVES AND OUR CELLS

How boys and girls grow into men and women and become physically mature

It is indeed an important time in the lives of boys and girls when they start becoming young men and women. In addition to the new faces and bodies they see in the mirror, they are beginning to have adult feelings. The days of yelling, playing pranks, and tearing around the school ground are over. Teen-agers become interested in their clothes and appearance. They grow conscious of one another's manners. They go out on dates. They often learn to drive the family car. And they start acquiring part-time jobs—from baby-sitting to clerking at the corner grocery.

Among the most noticeable changes in the adolescent's new self are those produced by the sex glands. Like other glands in the endocrine family, the sex glands secrete hormones. This change starts at puberty. For girls, puberty usually begins between the ages of twelve and fifteen, though it may be as early as ten or as late as seventeen. For boys, it is a little later—usually between thirteen and sixteen.

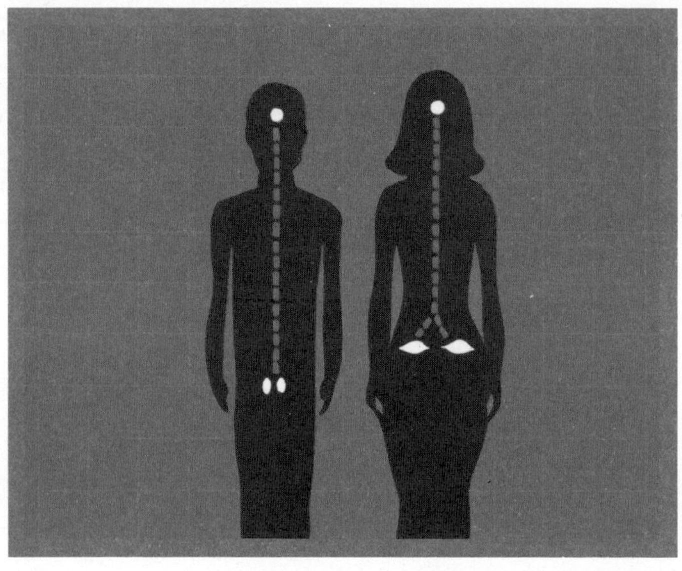

Hormones secreted by the pituitary gland influence the growth and function of the testes and ovaries. The testes, when acted upon by the pituitary, begin to secrete the male sex hormone, and the ovaries the female sex hormone.

As we already know, a special hormone from the pituitary gland sends a message to the sex glands. This is the signal that starts the boy's testes and the girl's ovaries making hormones of their own. In turn, these sex hormones travel up to the pituitary gland. They act upon this gland in such a way that it secretes less of the growth hormone. Thus growth slows up and finally stops completely.

In boys, the hormones secreted by the testes promote the growth of the male sex organs. And other changes take place. The boy starts sprouting whiskers. Hair grows on the legs, on and under the arms, in the pubic region, and sometimes on the chest. This is all perfectly natural. In fact, the word *puberty* comes from the Latin verb *pubescere,* meaning "to become hairy."

The beard grows in an interesting way. First, the soft hairs above the boy's upper lip become more pronounced and darker in color. If allowed to grow, they would form a straggly mustache. Next, the hair on the cheeks shows a similar change—soon to be followed by the appearance of "sideburns" down the sides of his face, the lower part of the chin, and immediately in front of the ears. Finally, the hair on the throat becomes coarser and longer. By this time the boy has started to shave. The thickness and color of the beard varies greatly among grown men.

The sex hormones also cause the boy's voice to change. A child's voice is high, often shrill. At adolescence it starts getting deeper. As the voice box grows in size and the vocal cords in length, a boy's voice "cracks"—often in the middle of a sentence. Soon, however, his speech assumes a new, masculine tone. Adult male voices range all the way from tenor to husky bass.

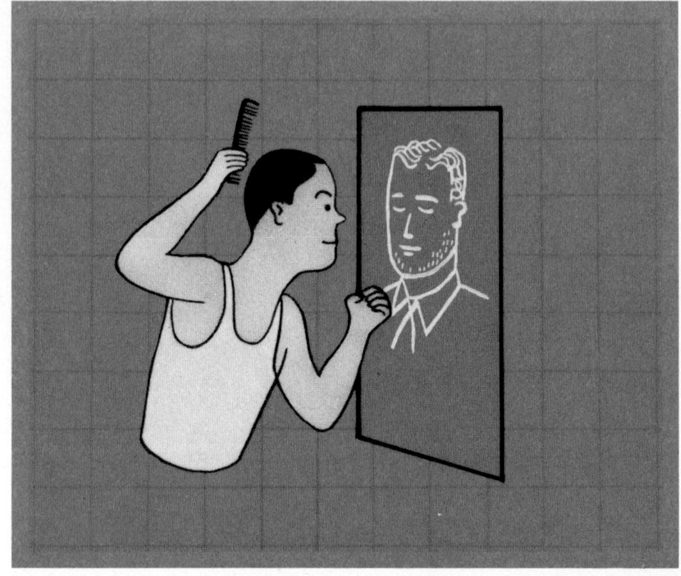

The male sex hormone causes the voice to get deeper and whiskers to appear on the face. It also causes hair to grow under the arms, in the pubic region, and elsewhere on the body. The physical changes usually make the boy feel more manly.

Meanwhile, the shape of the boy's body becomes more like that of a man. He grows more muscular, with broad shoulders and narrow hips. By high school graduation or before, he'll be wearing a man's suit to show that he has almost grown up.

As the girl approaches puberty, the hormones from her ovaries also affect her appearance. Like the boy, she finds hair growing under the arms. Hair also grows in the

The female sex hormone causes the girl's voice to deepen a little, hair to grow on the body, and the breasts to develop. These changes usually make the girl feel more womanly.

pubic region, usually in a triangle-like patch straight across the top—a different shape from that of boys.

Hair grows thicker on the arms and legs. Sometimes, especially among girls with dark hair and skin, the fine down on the sides of the face and upper lip becomes more noticeable. This doesn't mean, however, that brunettes are any less feminine than blondes—body hair on the latter just doesn't show as much.

Although the girl's voice also changes as she becomes mature, the change takes place more gradually than in boys and is hardly noticed. Feminine voices become richer and fuller, and range from deep alto to high soprano.

In the process of reaching sexual maturity, the girl's breasts develop. The mammary or milk-producing glands, located in the breasts, become larger, so that when the girl marries and becomes a mother, she can nurse her baby. When the breasts start to grow, they may feel tender or sensitive for a time. At this age, girls start wearing brassieres and taking a new interest in clothes. Some girls may be barely in junior high when their breasts start to grow, while others may be well into high school. Some will reach maturity with a small bust, others with a large one. These are all normal differences.

Other parts of the girl's body change, too. Her sexual organs grow. With the rounding of her hips, she assumes a more feminine appearance. As she approaches womanhood, she likes to experiment with a touch of lipstick or fresh-smelling cologne. She is also likely to spend a longer time combing her hair. All this attention to how one looks is another stage in the process of reaching adulthood.

The testes and ovaries have a second function, in addition to secreting hormones. As boys and girls become sexually mature, these glands produce male and female sex cells. These very important cells provide the beginning of new life. It is these cells which are responsible for human reproduction.

The male cell is known as the sperm cell. Mature cells are first produced at puberty. The sperm cell is unbelievably tiny, only one five-hundredths of an inch in size. It cannot be seen without the aid of a microscope. If it were magnified many times, the sperm cell would look something like a tadpole. It has a head, containing the nucleus, and a long threadlike tail. The nucleus is the tiny round spot in the center of the cell.

The female cell, known as the ovum or egg cell, is much larger than the sperm. Although the largest single cell in the human body, the ovum is nevertheless smaller than a pinpoint. The outer edge or the cell wall surrounds the nucleus.

We spoke of the growth of the sex organs (also called genitals) as an important development during puberty. In childhood, before this period is reached, the genitals are only about one-tenth of their adult size.

The girl's sex organs are, for the most part, inside the body. In addition to the sex glands or ovaries, the main parts include the vulva, the vagina, the uterus, and the tubes.

The outside portion of the girl's sex organs, between her legs, is called the vulva. It consists of thick folds of skin known as the outer lips, with tissue underneath. A groove runs between these folds. Inside the outer lips there are thinner folds of tissue called the

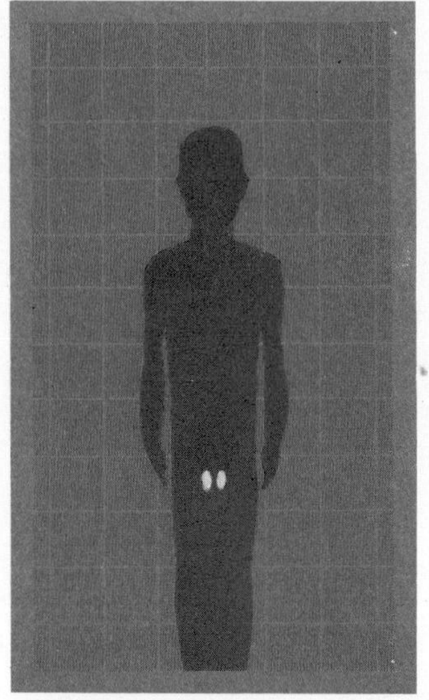

The testes produce the male sex cells (sperm) which are so small that they can be seen only with a microscope. Each sperm cell has a head, containing a nucleus, and a long thread-like tail that wiggles and causes the cell to move.

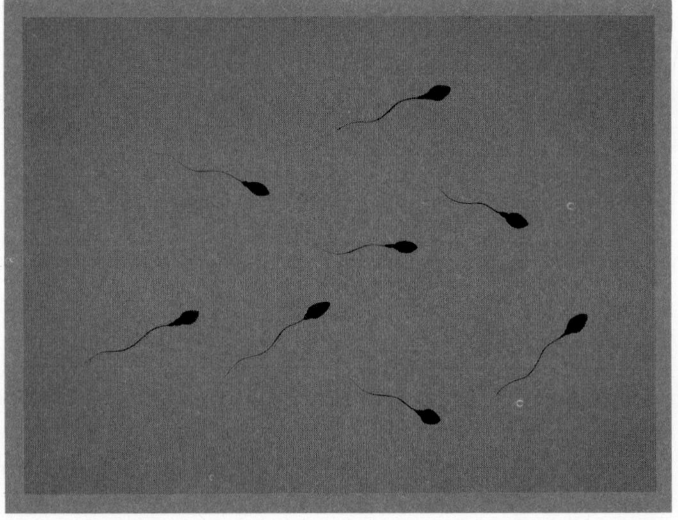

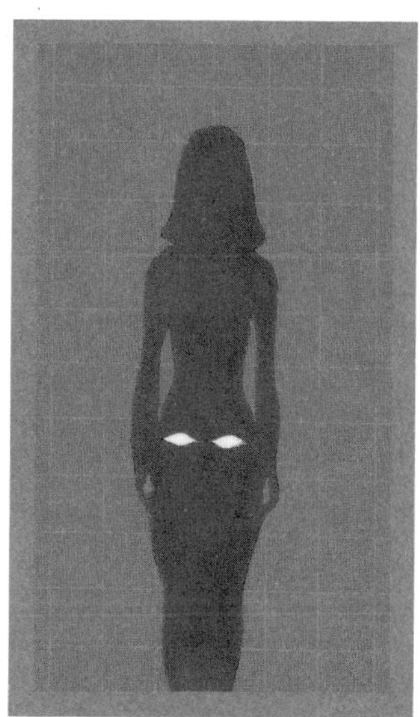

The ovaries produce the female sex cells (ova) which are no larger than a pin point. Each ovum has a round nucleus inside.

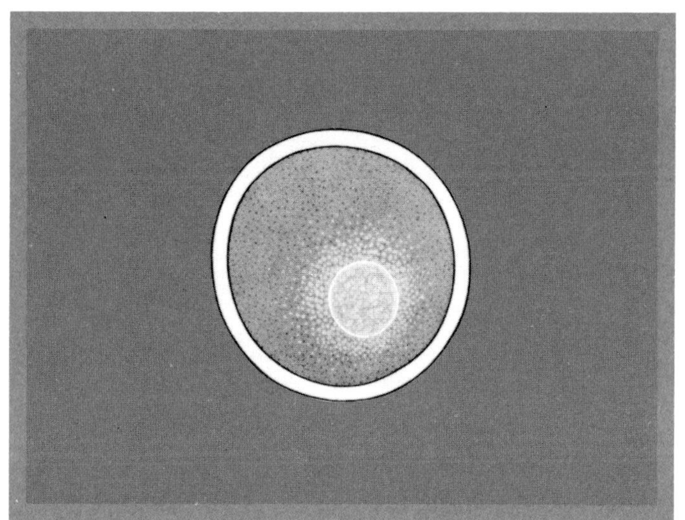

inner lips. These folds serve to protect the two body openings in this region. One opening, at the top of the vulva, is used to discharge urine (the body's waste fluid). A little farther back is the opening of the vagina. The

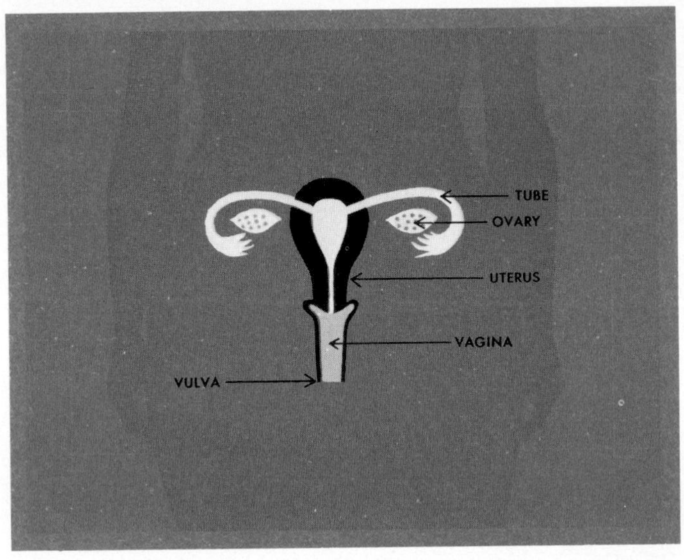

The girl's sex organs consist of the ovaries, tubes, uterus, and vagina, all inside the body, and the vulva or outside part.

vagina is a canal about four inches long, extending upward and backward inside the body. Its walls are composed of stretchy muscle fiber with a rough texture.

The upper end of the vagina joins the uterus, which is hollow and pear-shaped, with elastic muscle walls.

This is where the baby starts life inside its mother. Being elastic, the uterus can stretch many times in size to make room for a growing child. Opening into the upper part of the uterus are two tubes, one at each side. Each tube extends for two or three inches and curves partly around an ovary. These tubes are delicate passages with hairs in the lining.

The two ovaries are thin and oval-shaped—one to two inches long, about an inch wide, and one-fourth of an inch thick. Located on the right and left of the uterus, they serve the same function in the female as the testes in the male—that of producing hormones and sex cells. Within the ovaries are many small, rounded chambers. These contain great numbers of immature egg cells. A single egg cell is called an ovum, and many egg cells are called ova. When a girl reaches puberty, one of the chambers opens up, releasing a mature or ripe egg. This procedure is known as ovulation. It usually occurs once a month. The ovaries take turns in releasing egg cells. Although the ovaries contain thousands of cells, only three or four hundred ripen within the average woman's lifetime.

After leaving the ovary, the egg cell is free for a short time in the abdominal cavity near the open, funnel-shaped tube. With the help of the tube's fragile, fringelike ends, it is caught and drawn in. Then, aided by wavelike movements of the hairs in the tube and by the muscles of the tube itself, the ovum starts toward the uterus. This trip takes several days.

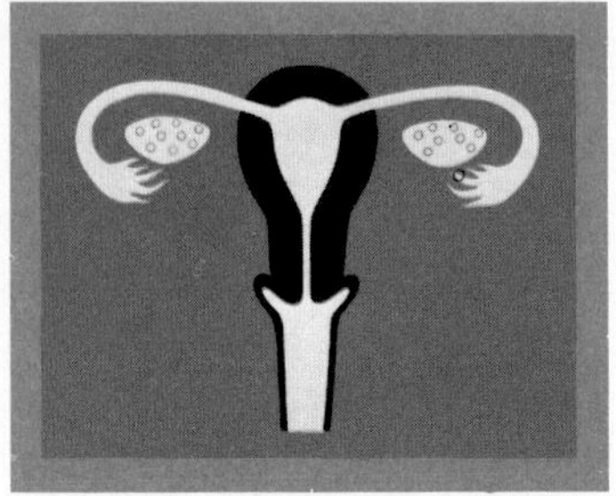

From adolescence on, an ovum ripens and leaves the ovary about once a month.

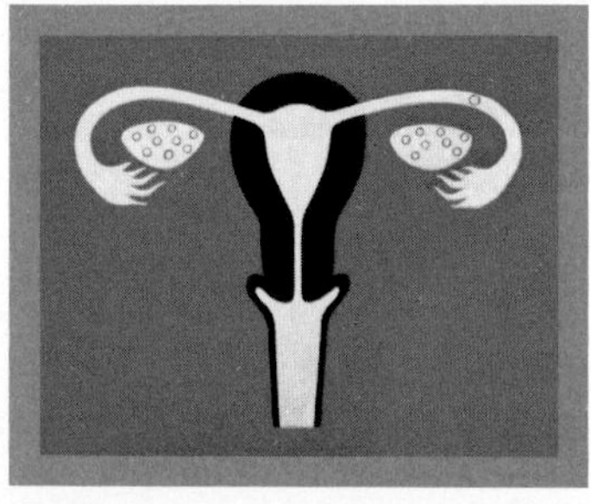

The tiny ovum, coming from the ovary, is drawn into the tube.

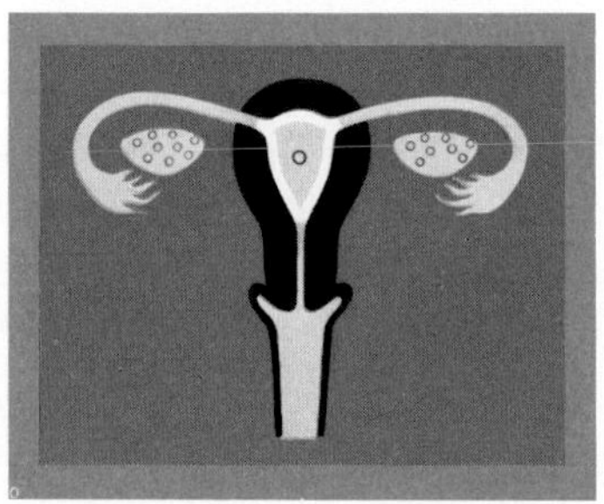

As the egg moves along, the uterus, over a period of days, builds up an inner lining richly supplied with blood.

Meanwhile, the uterus is being supplied with extra blood and tissue in case this material is needed to nourish a new life. If fertilization occurs—which means that a sperm has joined an egg—human growth begins. But if the ovum fails to meet a sperm cell, no new life starts and the egg cell breaks up, and since the uterus no longer needs such a thick lining, the innermost layer of tissue peels off. Along with some blood, it leaves the body through the vagina. This discharge is called menstruation, and lasts as a rule from two to seven days. The word menstruation comes from the Latin word *mensis,* meaning month—the menstrual periods taking place monthly.

As we know, the process of ovulation also occurs monthly—about two weeks after menstruation. Ovulation and menstruation mean different things. During ovulation an egg leaves the ovary and travels down the tube to the uterus. If the egg is not fertilized, menstruation follows. The uterus gets rid of its extra lining and blood. After menstruation the uterus again starts to build up blood and tissue, another ovum is released, and the whole cycle repeats itself. This process continues month after month, year after year. It is interrupted only when an ovum becomes fertilized and attaches itself to the uterus. This condition is known as pregnancy.

Thus, as they approach maturity, girls undergo a new and eventful experience—they start to menstruate. This begins at the time of puberty, which was discussed earlier. For some girls, the first menstrual period occurs when they are no more than ten or eleven years old. For

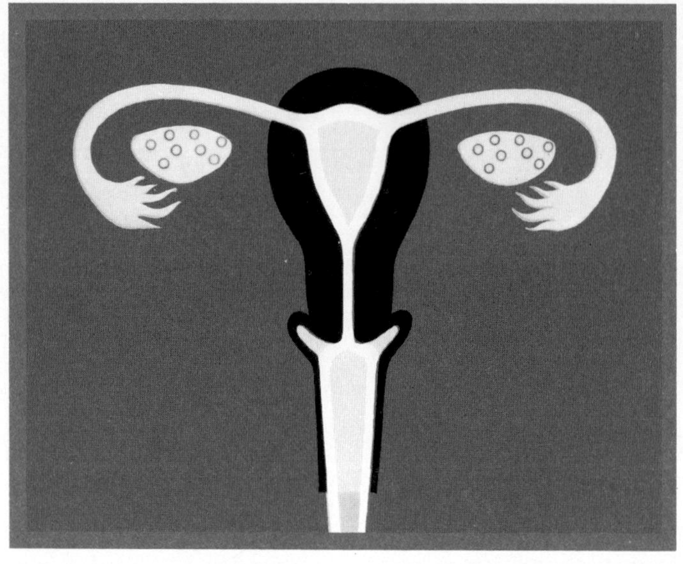

If the ovum does not meet a sperm cell, it dies and disappears. The uterus then sheds its inner lining along with some blood. This process is called menstruation.

others, it may come as late as sixteen or seventeen. Menstruation is a perfectly normal function of the body. It means that a girl is becoming a woman. During menstruation sanitary napkins are worn to protect the clothing.

At first, the menstrual periods are likely to be irregular. One month, the period may begin several days early; another month, it may be a little late. It may, in some cases, skip a month. Within a year or two, however, a regular rhythm is usually established.

The time between menstrual periods is known as a cycle. These cycles vary from woman to woman. Although the average cycle lasts twenty-eight days, menstruation seldom occurs like clockwork. A woman with a twenty-eight-day schedule will probably menstruate within twenty-six days one month, within twenty-seven or twenty-eight days another, and sometimes not for thirty days.

A girl or woman may have a regular cycle as short as twenty-one days. For others, the time between one period and the next may be as long as thirty-five days. The length of the menstrual period also varies. It can take from two to seven days. Usually the blood flow lasts four or five days. The flow is heavy early in the period, especially during the second or third day. Then it gradually decreases and finally stops.

Menstruation is a completely natural process, during which life should proceed as usual. The girl's fun and activities need not be hampered. Some physical exercise on the first day or two often helps the body to function

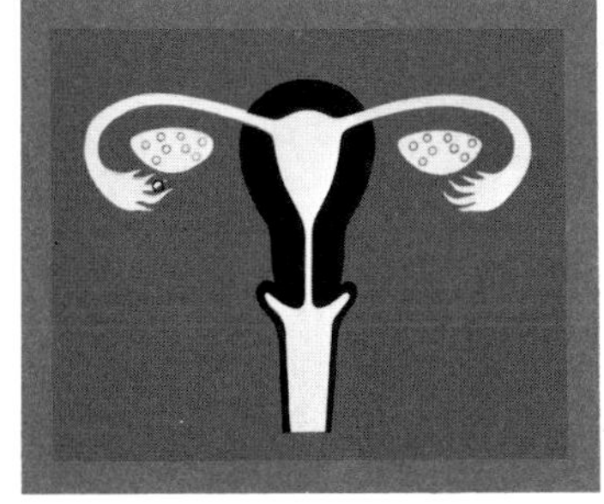

In a couple of weeks, a new ovum is released, usually from the opposite ovary, and unless the egg is fertilized, the menstrual cycle is repeated.

even more smoothly. A girl can continue to dance and take part in sports, avoiding only the most strenuous exercise—since she should not get overtired.

During menstruation the body is more sensitive to extremes of temperature. For this reason, girls are advised to avoid extremes of heat and cold. At such times, however, a daily warm bath becomes even more necessary than usual for the sake of cleanliness.

While most women and girls feel perfectly well during menstruation, some feel uncomfortable and out of sorts. A few girls suffer from painful cramps early in the period. Severe pains are often a sign that the body needs attention, and in these cases a doctor's advice should be sought. As a rule, however, there is no reason why a girl should expect to be laid up or pampered during her menstrual period. Such names for menstruation as "the curse" and "the monthly sickness" are misleading and out of date. Actually, regular menstrual periods are a sign of good health.

At some time usually in their middle forties, women go through the period of the menopause or "change of life." After becoming increasingly irregular, menstruation stops, and the ovaries no longer produce mature egg cells. The sex hormones which these glands started secreting at puberty are produced in smaller quantities. This means the woman can no longer bear children.

Although most women experience this change in their middle forties, the time varies from person to person. Both during and after the menopause, everyday life continues as before, and the woman is usually her normal self. Occasionally, however, women become worried and upset by the change, which may take place over a period of several years. Especially at this time in their lives, women need love and understanding from other people, particularly from those in the family.

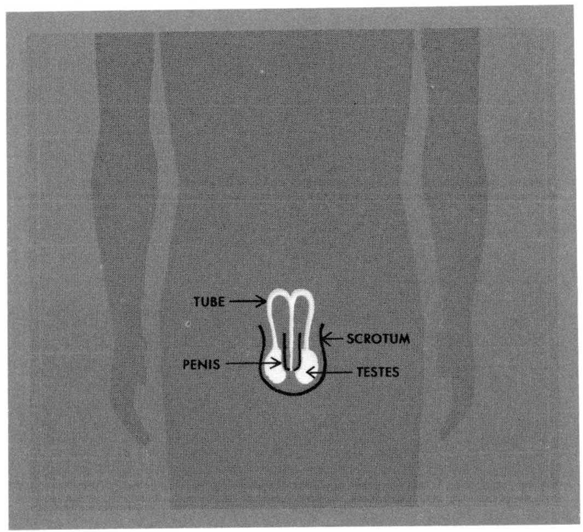

The boy's sex organs include the tubes, inside the body. The penis and scrotum, which contains the testes, are on the outside.

In the male, the sex organs consist of the main tubes, inside the body, and the penis and testes, outside the body. The testes (or testicles) are contained in a pouch of loose skin called the scrotum, which hangs immediately behind the penis. The reason it is located outside the body is that the stored-up sperm cells cannot withstand the higher temperature within the body. One testis usually hangs a little below the other. In warm weather, the testes hang lower in order to protect the sperms from the body's heat. In cooler weather, muscles draw the testes closer to the body.

In most boys, both testes descend to their place inside the scrotum at about the time of birth. Now and then one testis fails to come down and remains inside the body, but this condition can usually be corrected by a doctor. If both testes are allowed to remain inside the body, conditions are not normal for the making of sperm cells. This means that the boy will become sterile—in other words, he cannot become a father.

The testes are egg-shaped, approximately one and three-fourths inches long and an inch thick. Each testis is composed of two or three hundred tiny chambers, or sections, separated by partitions. And each section is full of threadlike, tightly coiled tubes. It is within these little tubes, or tubules, that sperm cells are made. One testis alone contains about eight hundred tubules, each two feet long.

Behind each testis lies a storing place for the sperm cells. This storing place is also outside the body, in the scrotum. It looks like a mass of tiny tubes. It is here that the sperm cells are kept, millions at a time, after being manufactured in the testis. Upon signals from the nervous system, the sperm passes through the tubes and out of the penis.

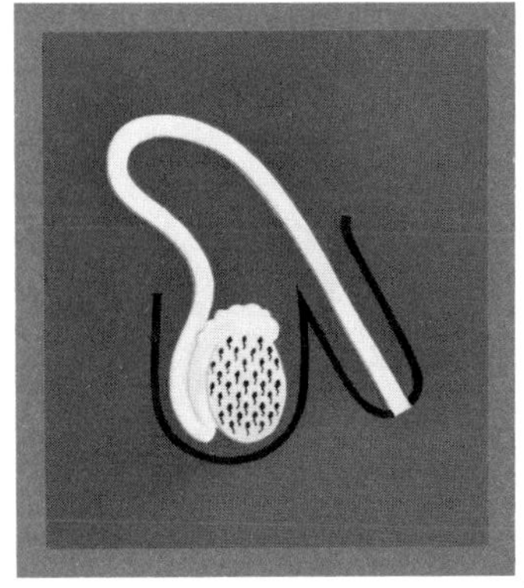

Sperm cells grow and multiply, and are stored in the testes.

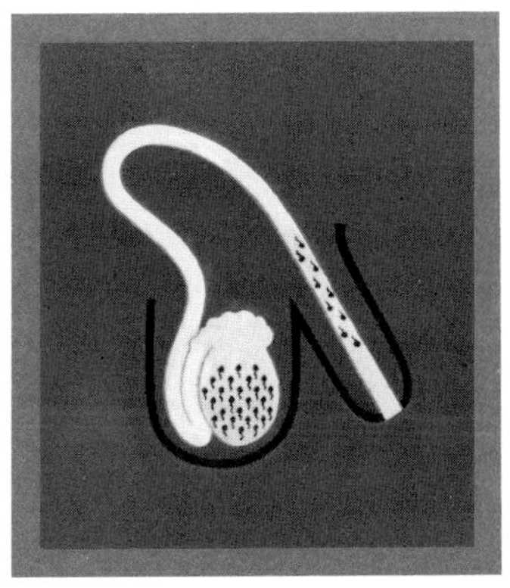

During mating and sometimes during sleep, the sperm cells pass from the testes through the tubes and out the penis.

The place where the sperm cells are stored is connected with a larger tube about sixteen inches long. This is called the main sperm duct. Each testis has its own main duct, which enters the lower abdomen and curves under the bladder. These two ducts, one from each testis, enter the single channel which leads out of the penis.

Although this same channel through the penis carries both urine and the sperm-carrying fluid, they never leave the body at the same time. Through the action of certain nerves, the tube carrying urine to the penis is shut off whenever sperm cells are to pass through.

Several glands lie along the pathway of the main tubes. Some of these glands produce a white fluid called semen, in which the sperms travel.

The penis is located in front of the body, at the base of the abdomen. Like other parts of the body, it varies in size from person to person. In many boys, the cap or foreskin covering the sensitive tip of the penis may be absent. This is due to an operation called circumcision, which is performed on some male babies shortly after birth. Most doctors believe that this aids cleanliness.

The penis usually hangs down, limp and soft. At certain times, however, blood rushes into the penis. Then the tissues swell and the penis becomes hard, sticking up and out from the body at an angle. This condition is called an erection. Before sperm cells can leave the body —during mating and sometimes during sleep—the penis must be in a state of erection.

Once in a while the penis becomes erect when the boy is asleep. A small quantity of fluid spurts out, leaving a sticky substance on the sheet or pajamas. This is known as a seminal emission, sometimes called a "wet dream" because a dream accompanies the event. It is a sign of sexual maturity and usually occurs in boys for the first time during adolescence. Unlike urinating, an emission takes place automatically, without warning. It is a natural function that occurs from time to time in the life of nearly every boy.

It is also not unusual for boys and men to awake in the morning with the penis erect. This may be caused by the bladder being full, in which case it presses on certain glands. Daytime erections can also occur, but they are seldom noticed except by the boy himself.

In later life, usually in their fifties or sixties, some men pass through a period which is somewhat like the menopause in women. During these years the glands which produce sex hormones become less active. When this period is over, men are less likely to become fathers than when they were in the prime of life.

Now let us see what questions boys and girls ask about physical development and the differences between men and women.

Why can't some men become fathers?

There are several reasons. A man is said to be sterile if he cannot become a father. Sterility may be caused by damage to the male testes, or by ill health, which can prevent the production of sperm cells. Sometimes an infection blocks the tubes and prevents sperms from leaving the body. A man can also be too old for fatherhood, though some men are still able to reproduce in their seventies. Occasionally men of eighty have become fathers.

How do sperm cells get out?

After nerve signals have caused the penis to become stiff and erect, muscular contractions squeeze the sperms, which, combined with a milky fluid, semen, move into the tubes and finally out of the penis.

When boys lose their sperms, is that like menstruation?

No. Seminal emissions, as they are called, are not regular events that take place once a month. Now and then during sleep the penis becomes erect and a little semen, carrying the sperm cells, leaves the body. This event lasts no more than a minute. Boys don't expect these emissions or "wet dreams" at any certain time. And, since only a small amount of fluid is discharged, they do not wear anything to protect their clothing.

When should a boy begin to shave?

Whenever his beard starts to look unsightly. The time at which the beard begins to show varies widely from one boy to another, depending on the secretions of the sex glands.

How can a girl improve her appearance if she has too much hair on her face or on her arms and legs?

There are various ways of removing superfluous hair from the face and limbs. Hair can be plucked, cut, or even shaved—but before attempting any of these remedies, it is very important for a girl to consult her parents, the health teacher at school, or the family doctor, who can suggest the best ways to remedy unattractive facial or body hair.

Why do women have bigger breasts than men?

The breasts of women contain mammary glands, which supply milk to newborn babies. A girl's breasts fill out during adolescence so that when she marries she can become a mother. A man's chest does not develop because it does not have to produce milk. The secretions of the sex glands play an important part in the development of a woman's breasts.

What is meant by the womb?

The womb is another name for the uterus. It is the home of a baby inside its mother's body. During the baby's nine months of growth, the uterus stretches and becomes much larger. When not carrying a baby, the uterus is hollow and shaped like a pear. It is connected with the outside of the body by the vagina.

Is the uterus located under the intestines or near the back?

The uterus is located beneath and in front of the intestines, in the lower part of the abdomen. The abdomen contains the stomach, intestines, internal sex organs, and certain glands.

Where is the baby born out of a woman's body?

There is a special place in the woman's body for the baby to come out. This birth canal has its opening between the legs. It is called the vagina. The vagina also is the place where the male sperm cells are deposited. And it is through this same canal that the menstrual flow leaves the body each month.

Is menstruation absolutely necessary?

Yes, because each month a woman's body is preparing for a possible birth. First, the uterus builds up a rich inner lining to provide nourishment for a new life. If fertilization does not take place, however, the extra tissue is not needed. It is shed and leaves the body along with some blood and the unused ovum. This process is repeated again and again unless an ovum is fertilized.

Old women and very young girls do not menstruate, and therefore cannot have babies.

Is it harmful if the menstrual flow is delayed?

Ordinarily, no. In early adolescence, a girl's menstrual periods are usually irregular. The flow is naturally a few days early or late. But if menstruation is delayed two weeks or more, after having been regular, then the girl should consult a doctor.

Is something the matter if girls menstruate for two weeks?

A doctor should be consulted to answer that question. Most girls menstruate from two to seven days. Two weeks would be very unusual. This condition might be a warning signal that the glands aren't working as they should, or that something else is wrong.

Can girls take baths while they are menstruating?

Certainly. To keep the body clean at this time is even more important than at others. But very hot baths and very cold showers should be avoided. Extreme temperatures can either increase or slow down the menstrual flow.

What causes cramps?

Several things can cause cramps. The glands may not be producing a sufficient quantity of certain hormones. Or something may be wrong with the position or structure of the uterus. Nervous upsets can cause cramps. If menstruation is painful, month after month, the body may be suffering from an injury, an infection, or high blood pressure. A doctor should be consulted. It is always better to ask the advice of a doctor, rather than take patent medicines and pills that haven't been recommended by him.

How old are most women when they stop menstruating?

Menstruation stops for most women in the middle forties. This change comes about gradually and is called the menopause or "change of life." After the menopause, a woman can no longer become a mother. The ovaries stop producing egg cells, and the manufacture of female sex hormones slows down.

Why do all women stop menstruating when they get old?

This happens because of a change in the glands of the body, called the menopause. The ovaries slow down their work of making eggs and finally stop making them altogether. Since there are no eggs to be fertilized, the uterus no longer needs to prepare nourishment for a baby. The lining is not shed each month, and there is no flow of blood.

CHAPTER THREE

PREVIEW TO PARENTHOOD

How a new life begins — another step in the cycle of human growth

THUS FAR, we have considered two phases of maturity: first, physical maturity, or full growth of the body; and second, sexual maturity, or full growth of the reproductive organs. Sexual maturity is accompanied by changes in the shape of the body. Young people differ greatly in the ages at which they reach these milestones along the road of growth. Many boys and girls grow tall or heavy before their bodies assume an adult look, or before they are sexually mature.

But there are other kinds of maturity besides the physical and sexual. Before young people reach adulthood, they should have completed their education. For some, this will mean continuing their studies after graduation from high school. They may enter a business school, a college, or train elsewhere for a particular career.

Again, young people have to face the problem of choosing a profession. Before he can support a wife and family, a man must have a steady job. For both husband

Parenthood, in a broad sense, means more than having children. It means finishing school, finding a job, marriage, and planning the home together.

and wife, the bearing and raising of children means accepting responsibility.

In our society, men and women have usually reached their twenties before they are ready to marry and set up their own homes. By this age they should be fully adult—in their thoughts, feelings, and actions as well as in their looks. Adulthood—and with it, parenthood—opens up a whole new world.

Now, let us consider once more the important cells which enable men and women to become parents. Before a human being can start to grow, a sperm cell from the father must unite with an ovum inside the mother's body. Thus both father and mother are needed to start the life process.

A father and a mother care for each other in a very special way. One reason a man and woman marry is so that they can be together as much as possible. Marriage offers them a way to share their lives and to show their love for each other. Each partner tries to do things which will please the other, and a feeling of joy and contentment fills the home.

One of the important parts of marriage is to display affection by tender words and actions. Married couples enjoy sharing their thoughts and feelings. This closeness and understanding is expressed in the act of mating, by which the father and the mother become parents.

Mating or sexual intercourse occurs when the penis enters the vagina. The physical attraction of the woman's body for the man, and of the man's body for the woman, combined with the love they feel for one another, sends an increased supply of blood into the sex organs. This causes an erection in the man.

These preparations for the sexual act are aided by love-making and caresses. As the partners lie in bed in a close embrace, the woman's vagina releases secretions which serve to lubricate the vagina as the erect penis fits into its opening. The sensations of love and excitement

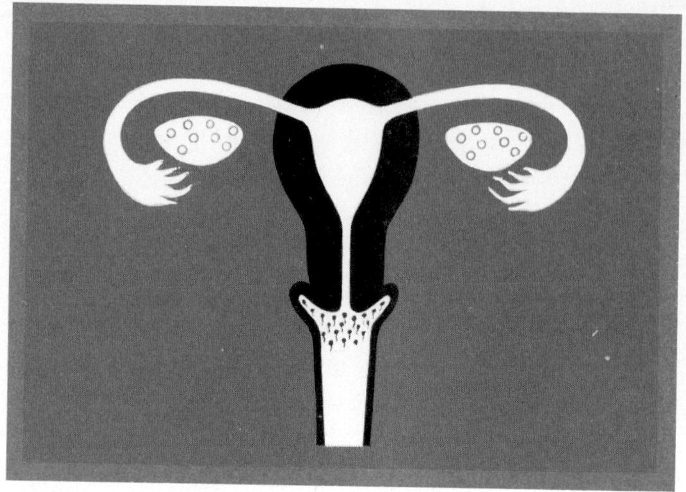

Human growth can begin only when the sperm cells of the father, during mating, pass from the penis into the vagina of the mother.

mount until they finally reach a climax. At the moment of climax, semen passes from the penis into the upper part of the vagina. Immediately after the climax is over, the man and woman feel a deep sense of peace and comfort.

Although only a small quantity of fluid leaves the father's body, it contains millions of sperm cells. The tails of these cells whip back and forth, propelling the tiny sperms up from the vagina and into the uterus. They then wriggle their way the length of the uterus and into the slender tubes. This trip takes from one to several hours.

If a live egg cell is in one tube when the sperm cells enter, fertilization will probably occur. When it does occur, this means that one sperm cell has succeeded in entering the ovum and a new life has begun. Not until a month later, however, can anyone tell whether or not fertilization has taken place. The first sign to warn the mother that a baby may have started to grow inside her body comes when her menstruation stops or is very slight. Should this happen, she ought then to consult her doctor in order to make sure that a new life has begun. If the doctor finds that new life has begun, the mother will soon experience some odd feelings in her stomach and a slight sensation of nausea or sickness. Such feelings are perfectly normal and do not last long.

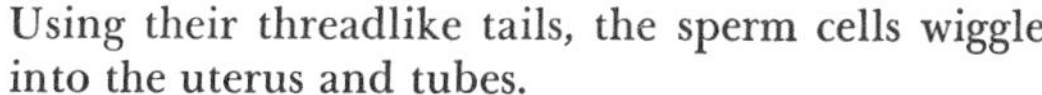

Using their threadlike tails, the sperm cells wiggle into the uterus and tubes.

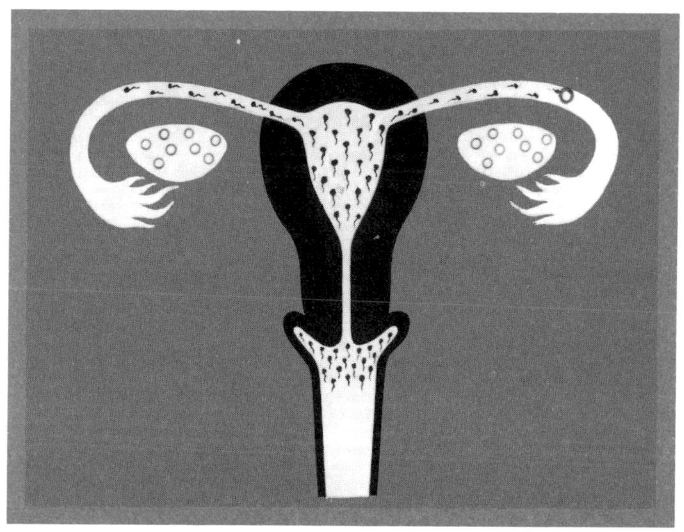

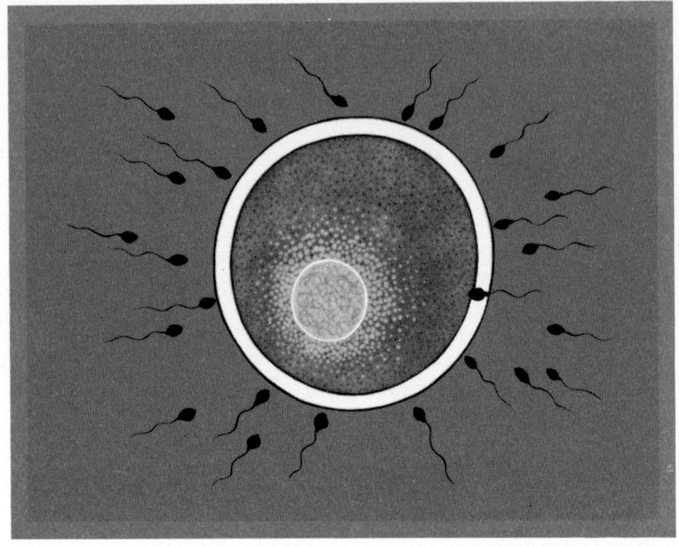

Upon finding an ovum, many sperm cells try to enter the egg, but usually only one succeeds.

Before proceeding further, let us remember that a new human being can start only from the union of a sperm with an egg. These two cells must join in order to start the cycle of reproduction.

In the process of fertilization, thousands of sperms collect around the cell wall of the ovum, in an effort to break through. This helps to weaken the cell wall so that one sperm—but only one—can enter. The others soon die, for sperm cells which do not meet eggs can live inside the mother's body only a short period of time.

It may seem strange that, though only one is needed

to fertilize the ovum, the male has millions of sperm cells. But now we can understand why. Unless the sperms are healthy and active, they cannot reach the ovum. Many are lost during the long journey through the vagina, uterus, and tubes. And many are needed to strike and weaken the cell wall of the egg so that one sperm can break through.

Now, let us see what happens to the lone sperm which enters the ovum. It buries its head in the ripe egg, and the tail drops off. The nucleus, or center, which is contained in the head of the sperm, grows larger and joins the nucleus of the egg.

The sperm cells on the outside of the egg die and disappear. The one on the inside of the egg drops its tail, and its nucleus starts to join the nucleus of the egg.

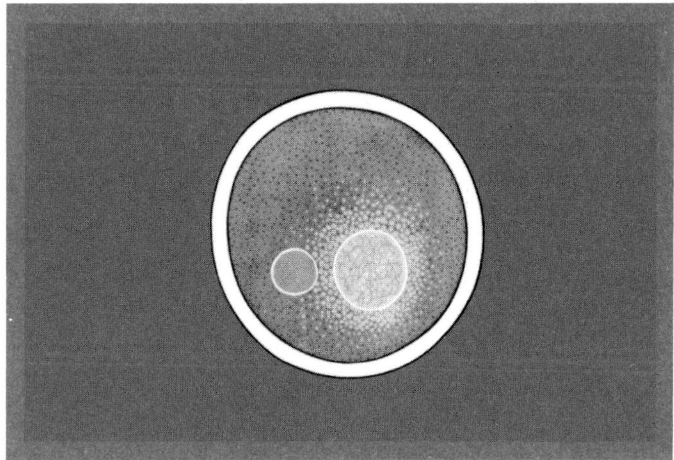

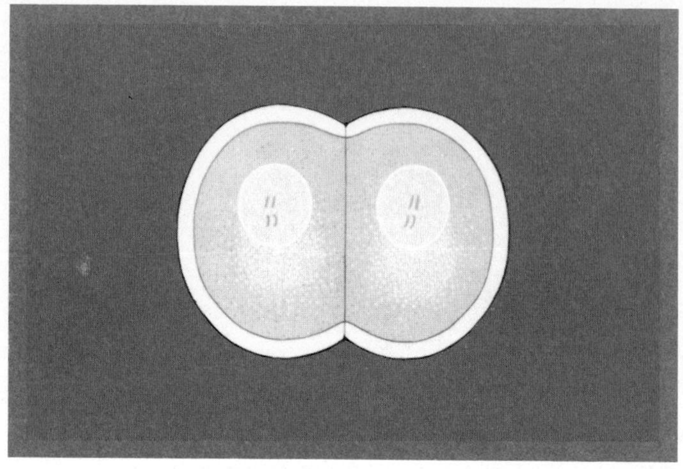

The egg begins to divide into several cells before it attaches itself to the uterus. Note the pairs of chromosomes in the egg. The ovum contains twenty-four pairs of chromosomes, but only two pairs are shown here.

The fertilized ovum then begins to divide, while moving slowly down the tube into the uterus. First it becomes two cells, then four, eight, sixteen, and so on. Soon it looks like a ball of cells. Some nine months later, when this "ball" of living matter has developed into a baby and is ready to be born, it will have more than two hundred billion cells!

A cell contains a nucleus—a disclike spot in the center—surrounded by a jelly-like substance which contains particles of food and chemical materials. On its outside is the protective cell wall.

The most important part of each sex cell is the nucleus. It contains tiny structures called chromosomes—meaning "colored bodies." These chromosomes carry chemical units which cause members of a family to resemble one another from one generation to the next. Chromosomes can be seen when the cell is stained a certain color and placed under a very high-powered microscope.

Most of the cells in the human body—nerve cells, muscle cells, bone cells, and blood cells—contain forty-eight chromosomes each. But the sex cells (the sperm and the ovum) contain only twenty-four chromosomes apiece. There is a very good reason for this. Remember that these two cells combine into a single new cell, which grows into a baby. With twenty-four chromosomes from

The chromosomes carry the substance that causes a child to resemble his parents.

the sperm and twenty-four from the egg, the fertilized ovum has forty-eight chromosomes in its nucleus. Half of these are from the father and half from the mother. As the fertilized egg divides, the chromosomes also divide, so that each body cell in the new human being has the proper number of chromosomes—forty-eight.

The chromosomes in a cell differ from each other in size, shape, and make-up. Some are threadlike, some straight, and some bent like little boomerangs. They are made of thousands upon thousands of microscopic parts, called genes. These genes are arranged on the chromosomes like beads on a string. They are the "special agents" which produce traits such as eye and hair color, build, and many other differences in people. The genes also act upon cells to determine which part of the body the cells will become. The cells, in turn, develop into glands, bones, muscles, skin, and other body structures.

Thus these genes are the cause of a boy having curly hair like his father, or a girl having brown eyes like her mother. The genes make up a chromosome, the chromosome is in the nucleus, and the nucleus is in the cell. When certain traits are carried down from parents to children, through the genes, we say they are inherited.

Heredity is therefore very important to the story of body growth. Other conditions, however, are also important for growth—things like food, sleep, and the surroundings in which a person lives. Even the customs of society, the way we get along with other people, the way we think, and the experiences of life: all of these go into the making of an adult.

Before we leave our discussion of the egg and the sperm, and how they unite to form a new being, let us see what happens to produce twins. According to statistics, twins appear about once in eighty-five or ninety births.

You will recall that each month, in most women, an ovary releases one ovum, which then travels down the tube. Once in a while, however, both ovaries send out eggs at the same time, or one ovary sends out two eggs. This means that two ova are in the tubes, both waiting to meet sperm cells. If mating takes place and both egg cells are fertilized, twins result.

Two separate fertilized eggs develop into the kind of twins known as unlike, or fraternal, twins. They may both be the same sex, or one may be a boy and one a girl. These twins look no more alike than other brothers or sisters. They usually differ considerably in size and in hair and eye color, even though they are born immediately after one another.

When fraternal twins develop, each has its own separate sac in the mother. Each is also attached to the uterus at a different place by its cord.

Sometimes twins grow from a single egg. During the first division of the egg, the two halves may separate and develop into two babies. The resulting twins grow in the same sac in the uterus and their cords are attached to the uterus at the same spot. Single-egg twins are known as like or identical twins.

Identical twins are always the same sex—either both boys or both girls. These twins are often very hard to tell apart. Not only are they the same size, their features and coloring are the same.

Triplets may be caused either by the fertilization of three separate egg cells or by one fertilized cell dividing into several parts, each part being a complete and independent cell with its own embryo. Or they may be the result of a combination of the two. For instance, two egg cells may be fertilized, but one of these two may divide into two parts, thus making three or triplets.

Similarly, quadruplets may be caused by a single fertilized egg cell dividing into four parts, by two or three cells dividing into four, or even by four separate cells.

Births of two or more babies at one time are known as "multiple births."

QUESTIONS AND ANSWERS

Did the egg divide into five parts to make the Dionne quintuplets?

As a matter of fact, it is probable that the egg divided into six parts, one of which was lost early in pregnancy. Scientists are pretty well agreed, however, that these quintuplets came from a single fertilized egg, and are therefore said to be identical as described in the last chapter. All five of them resemble one another very closely; and of course they are all girls.

What about Siamese twins? Are they just like other twins? What were their parents like?

Siamese twins are identical twins which did not separate from each other completely. The fertilized ovum becomes a cell mass. This starts to divide but doesn't quite finish dividing. There remains an area where the babies are joined—part of the chest, abdomen, back, or side. If they live, they go through life fastened together, unless an operation can be performed to separate them. This is not an inherited tendency.

Very few Siamese twins are ever born. The name "Siamese twins" comes from two male twins of this type who were born long ago in Siam.

Are twins smaller because the egg divides?

Not necessarily, but probably, because they have not as much room to develop inside the mother's body, as a single baby. In order to hold twins, the mother's uterus has to stretch larger. Also, the mother's body must provide extra food materials for two babies. Neither twin receives quite as much nourishment as when the food is not shared.

For this reason twins and multiple births are usually born earlier and are smaller than single babies. They often arrive a month or two before the usual nine months.

Wouldn't it be awfully crowded in there, with two babies? Wouldn't they bump heads and kick each other?

With twins, conditions *are* rather crowded inside the uterus. But as a rule, they are a little smaller than average babies. And, like all unborn babies, twins are protected by fluid which cushions the bumps or jolts they may get. Further protection is given to fraternal twins by the special sac in which each baby grows.

What happens when more than one egg cell is fertilized?

If two ripe eggs are in the tubes and each is fertilized by

a sperm, twins develop. The two fertilized ova move into the uterus. There they attach themselves to places on the lining and develop separately. These twins are unlike, because they come from different egg-sperm combinations. They are called fraternal twins, and they look no more alike than other brothers and sisters.

Why is it that some eggs are fertilized and become boys, while others are girls?

The sex cells carry structures in the nucleus called chromosomes, which determine the sex of the child. If the chromosome from the father is of the Y-type, the fertilized ovum will be a boy. If it is of the X-type, the child will be a girl. In other words, it is the *sperm* cell which determines sex. Whether a sperm with a Y-chromosome or one with an X-chromosome will fertilize the ovum is purely a matter of chance. Nothing the mother does during pregnancy can determine the sex of the baby.

What makes the baby look like the mother or the father?

The substance that causes a child to look like his parents is contained in the nucleus of the sex cells. The chromosomes in the nucleus are made up of genes. These genes are tiny particles which determine build, features, eye, skin, hair color, and other traits connected with appearance and behavior.

How can you look like your father's mother, when your father doesn't look like her at all?

Family traits sometimes skip a generation. There are many possible combinations in the genes. A person may turn out to look like some other relative—not at all like his parents. Traits may appear, then skip a generation, then show up in the grandchildren.

Why do some people have red hair?

Hair color—black, red, brown, blond—is caused by pigment. Pigment is a coloring substance in the tissues and cells of the body, and is determined by the genes.

If sperms were put into the female when an egg had already been fertilized, what would happen?

Nothing would happen. If there is a fertilized ovum in the uterus, a message is sent to the pituitary gland. This gland sends out hormones which prevent the ovary from producing any more eggs. Therefore, there are no more eggs in the tubes to fertilize until after the baby is born.

How long do the sperms live?

Only a short time inside the female. Probably not more than a day or two. Those that do not fertilize an egg die.

What would happen if two sperms entered one egg?

The egg would not grow because of the abnormal number of chromosomes. However, two sperms very rarely enter a single egg. As soon as one sperm cell succeeds in uniting with the ovum, a change takes place in the cell wall of the egg. This prevents any other sperms from entering. It also acts as a natural safeguard, permitting only one sperm and one egg cell to join.

How long does the penis have to stay in the vagina to let loose the sperm cells?

Not very long. Perhaps only a few minutes. The time varies, depending on the man and the woman.

Does sexual intercourse always cause the female to become pregnant?

No, not always. The female does not become pregnant unless a sperm cell fertilizes an ovum. A woman usually produces just one ripe egg a month, and it may not happen to meet a sperm cell during intercourse. Also, either the man or woman may be sterile, in which case the ovum cannot be fertilized to produce a baby.

When does mating take place? During menstruation or not?

There is no special time for the mating process to take place.

How do sperm cells enter the vagina?

The sperm cells pass from the penis into the upper part of the vagina. This happens during mating, or intercourse. In mating, the penis becomes stiff and hard, so that it can fit naturally into the vagina.

Can everybody be parents?

No. Sometimes either the husband or the wife is physically unable to reproduce. This condition is known as sterility. These people may be otherwise healthy and living a normal life.

In addition, older people—or those past childbearing age—cannot become parents.

Why do some young men and women have children before they are married?

From time to time all human beings who have reached sexual maturity feel the urge to mate with the opposite sex. This urge can lead a young man and woman to mate before they are ready to marry and establish a home. As the result of such a union, a baby may be born. Since our society disapproves of unmarried parents, not only must these parents face the disapproval of society, their child must also face it. Moreover, because these unmarried parents are not prepared for the responsibilities of a home and family, they are not able to give their child the care and love all children need.

Why cannot a woman have a child so long as there are still eggs in her body?

A woman can be sterile—that is, unable to have children. Sterility can be caused by an illness or a disease which closes the tubes, preventing the sperms from reaching the ovum. Sometimes the woman is able to have a baby, but the husband is sterile. The egg cell is there, but it is never fertilized.

When a woman is so old that she cannot have a baby, is this because she has used up all the eggs?

No. Lots of unripened or immature eggs still remain in the woman's ovaries after menopause. But since they no longer mature, they cannot leave the ovary and move into the tubes. For this reason fertilization cannot take place.

Have old women ever been known to have babies?

That depends on what is meant by "old." A few women have been known to bear children in their early fifties. This means that the period of their menopause came unusually late. But it is doubtful if any women in their seventies or eighties ever have been mothers. Ordinarily, the late forties mark the end of the period in which a woman can have babies.

A story in a newspaper said a woman was going to be sterilized. What does this mean?

The woman had an operation performed which made it impossible for her to have a baby. This operation "ties off" the tubes, and prevents the sperm cells from reaching the egg cells. A similar operation on men prevents fertile sperm cells from leaving the testes.

CHAPTER FOUR

THE MIRACLE OF BIRTH

How the tiny human egg grows into a baby — and is born

COULD ANYTHING be more miraculous than that a speck of living matter much smaller than a pinhead should grow into a complete human baby? Yet that is what happens. The speed at which this speck grows within the woman's body is astonishing. During the nine months between fertilization and birth, the spot of living matter grows several billion times heavier. Yet between birth and adulthood, a baby increases in weight only twenty times.

Now let us return to the journey of the fertilized ovum—the combined egg-sperm cell. After traveling through the tube, the fertilized ovum enters the uterus and seeks a suitable place to lodge. During this trip, the cells continue to multiply and to arrange themselves in new ways. After about a week, the egg comes to rest. It embeds itself in the rich inner lining which the uterus has built up for this purpose.

The egg, after growing into a hollow ball of cells,

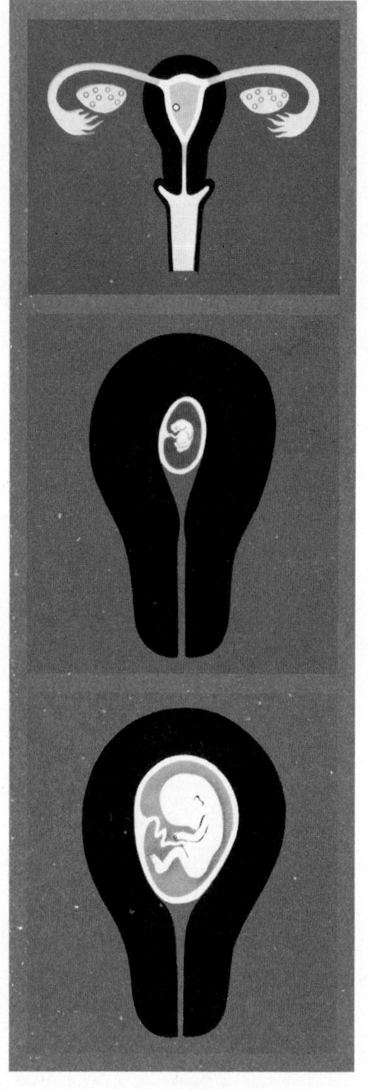

In the uterus, the fertilized egg attaches itself to the inner lining and begins to grow in size.

The egg grows and divides very rapidly, and in a month is an embryo like this.

At two months the embryo is much bigger, and is already beginning to look like a baby.

becomes longer and flattens out. Soon it is composed of three layers. It is from these layers that the different parts of the human body develop. The inner layer forms the digestive system, including stomach, liver, kidneys, and intestines. The middle layer provides bones, muscles, blood vessels, and blood. The outside layer of cells becomes the skin and the nervous system, which includes brain, spinal cord, and nerves.

This process by which the body is formed into its various parts is not at all simple. It involves many changes and is very complicated.

Part of the main cluster of cells develops into structures which house and feed the new life inside the uterus. Some of the cells form a sac which completely encloses the embryo. The growth of this sac accompanies that of the baby. The sac contains fluid in which the baby floats, and it is the child's home until he is born. The sac and its fluid together serve as protection, so that jolts to the mother will not injure the tiny being inside her body.

You may ask how a baby can live and breathe in water, as though it were a fish! The explanation is simple. The lungs are not used until after birth. Instead of breathing in oxygen from the air as we do, the unborn baby obtains this valuable substance from the blood of its mother.

In its earliest stage of development, the baby is called an embryo. At this stage, it bears almost no resemblance to a human being. When, after about three months, its main body parts appear, the embryo is called a fetus.

The embryo obtains its nourishment, as well as its oxygen, through a special structure called the placenta. The placenta is round, flat, and full of blood vessels. It develops on the wall of the uterus where the ovum first lodges.

The baby is connected to the placenta by a thick cord which contains the blood vessels of the baby only. This cord—known as the umbilical cord—is attached to the baby's body in the center of the abdomen. The place where it is fastened to the baby later becomes the navel.

The placenta acts as a filter between the blood of the mother and that of the embryo. From the mother's blood stream come food materials and other substances essential to the baby's life. These materials seep through thin membranes between the mother's blood vessels and the baby's blood vessels. The membranes are the placenta's filtering system. They sift out germs or impurities.

In turn, carbon dioxide and other waste products from the baby pass back through the same membranes into the mother's blood. The waste materials travel out of the uterus through the blood vessels. Eventually they are carried out of the mother's body along with her own waste products.

The remarkable thing about this exchange system is that the blood of the mother and the baby never actually mixes. Even while the embryo is growing in the warmth and shelter of the mother's uterus, it is already a separate being.

Now let us follow the growth of the embryo. After

one month, it is tiny and curled up, less than one-fourth of an inch in size. It has a knob for a head and, underneath this, a bump for the heart. It also has a tail! At this stage the human embryo looks like that of almost any animal.

At two months, the embryo is an inch long and weighs one-thirtieth of an ounce. By this time, its form is that of a human being. The tail has almost disappeared. The two-month embryo has a huge head, and its chin rests on a pot belly. It already has eyes, ears, nose, and mouth. There are tiny arms and legs, and the beginnings of fingers and toes. The heart is beating and sending blood through its body.

At the age of three months, the fetus (remember, we're giving it a new name) has grown much larger. Now it is between eight and ten times as heavy and almost four times as tall as it was four weeks ago. The fingers and toes have developed. Even so, the fetus is still no more than four inches long.

At four months, the baby-to-be has grown somewhat over six inches in length and weighs not quite four ounces. It has a bulging forehead, but its abdomen is not so large. Many of the bones are starting to harden. The fetus has a thin red covering for skin, through which the blood vessels can be seen. The muscles of its arms and legs can move. And the outside sex organs are apparent.

By the fifth month, the skin of the fetus is becoming thicker, and fine, downy hair covers its body. It is now about ten inches long and weighs up to eleven ounces.

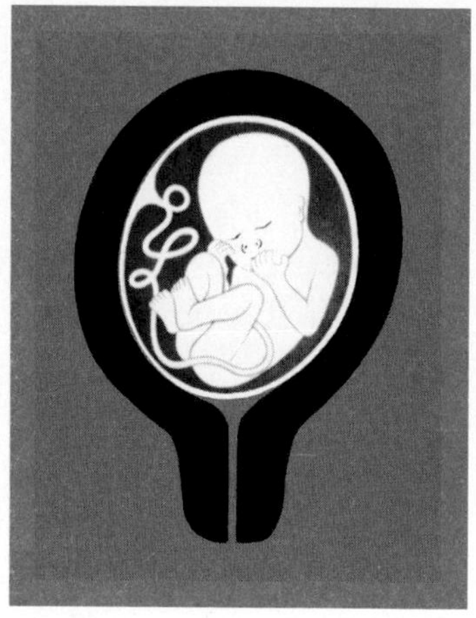

At four months the baby has arms and legs that move inside the mother.

It is at this period that the mother is aware of a new sensation. At first she senses only faint stirrings. Then, sometime later, she feels kicks, a sign that her baby is active. She is more conscious of these movements when sitting still than when she is moving about.

At six months, the fetus is lean in appearance, and about a foot long. It weighs close to one and one-fourth pounds and is beginning to look more and more like a newborn baby. Eyebrows and eyelashes have appeared. The organs inside the body, however, are still quite immature. A protective material, looking rather like lard, covers the body.

A month later, the fetus is still slender. It has settled in a head-down position, and hair has begun to grow on its head. It has gained a couple of inches in length and about one pound in weight.

During the last two months before birth, the soft down on the body of the fetus disappears; the skin becomes smooth and pink. The fetus spends some of the time moving and some resting. It grows plump and begins to store up fat. At birth, the baby is about twenty inches long and close to seven pounds in weight. The male usually weighs a little more than the female. Some babies are smaller than this average, some larger.

While these changes are taking place in the fetus, other changes occur in the mother's body. Since she has become pregnant, her menstrual periods have stopped. Long before the birth is expected, her breasts have begun preparing milk for the baby. As the mammary glands get ready for nursing, the breasts become larger and firmer. The nipples fill out and the area around them grows darker.

That part of the mother which changes most, of course, is her abdomen, which grows increasingly larger. The uterus, normally hard and pear-shaped, becomes softer. To make room for the growing fetus, it stretches to many times its usual size. As the uterus enlarges, the abdomen must also expand. At first, there is only a bulge in the lower part of the abdomen. This bulge continues to enlarge until, by the eighth month, the woman's abdomen has rounded out to the height of her ribs. A few weeks before the baby is to be born, the uterus gradually sinks down and the bulge grows even more prominent.

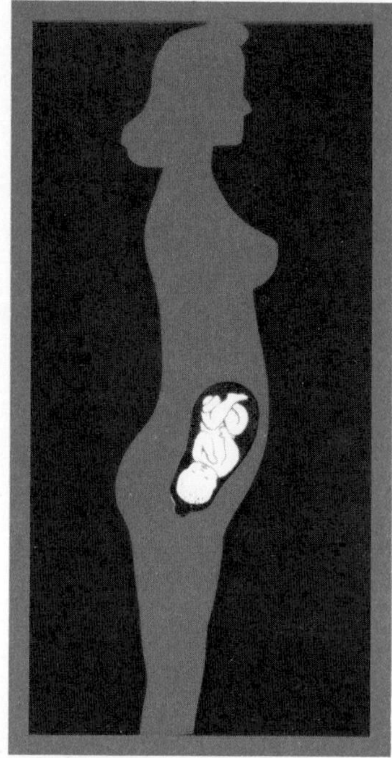

This is the position of the baby in the uterus inside the mother's body.

This is because the child is settling to a lower position in preparation for birth.

Throughout her pregnancy the mother pays regular visits to her doctor, who makes sure that her body is in normal working order. At this period, her health is particularly important. Since extra nourishment is needed for the baby, the doctor tells the expectant mother what and how much food she should take. She is careful to eat well-balanced meals rich in milk, fruits, and vegetables; and she drinks quantities of water.

The mother also watches her weight. She can safely gain as much as twenty-five pounds before birth. The baby, of course, weighs only a fraction of this, but extra tissues and fluids account for the added weight. Nights of sound sleep are also important for the expectant mother. If possible, she should try to have an afternoon nap. Exercise, too—particularly walking—is good for her.

While waiting for the new offspring to be born, most women feel perfectly well. Many women, in fact, experience an unusual sense of well-being during pregnancy.

As for the father, in seemingly minor yet actually important ways, he can be of great assistance to his wife during this exciting period. Paying her particular attention, he can see that she continues to live and enjoy a normal life. He can accompany her to the theater, the movies, to concerts and club meetings. He can help her in the kitchen and relieve her of some of the harder work about the house. And he can attend to her health and keep her in good spirits.

Finally, one fine day—about nine months after fertilization—the mother feels the beginning of mild cramps, rather like menstrual cramps. Sometimes they feel like a stomach-ache. Now she knows she has not long to wait.

These cramps are called "labor." They are caused by the muscles of the uterus starting to push the baby out. Labor is an apt name for these cramps, because the body has to work hard to force the child through the birth canal. At first the rhythmic pushes are felt about

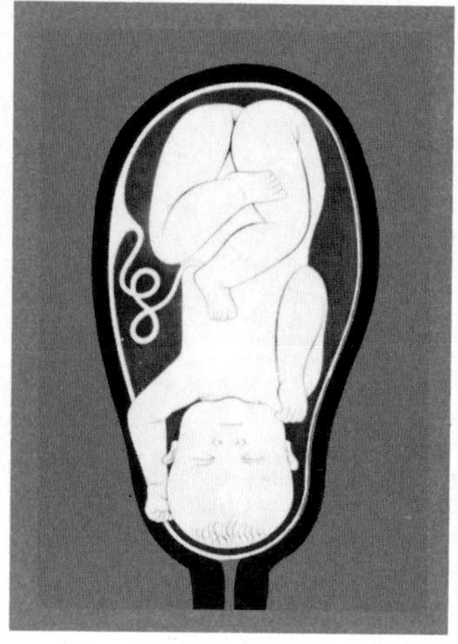

By six months the baby turns to this head-down position in preparation for birth.

If the mother were lying down, this would be the position of the baby just before birth.

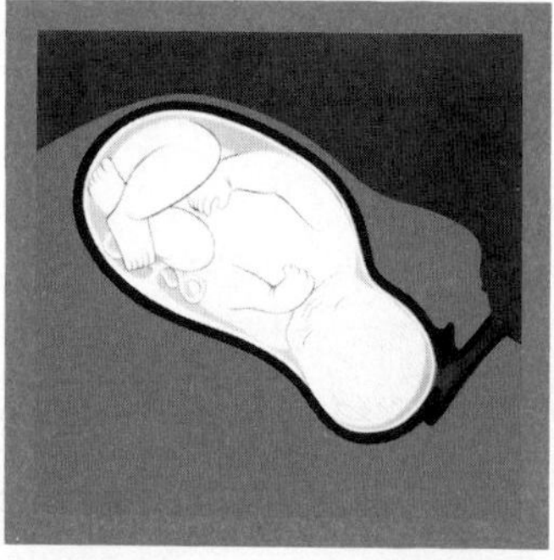

The baby's head settles to a lower position, and the muscles of the uterus begin to push the baby out.

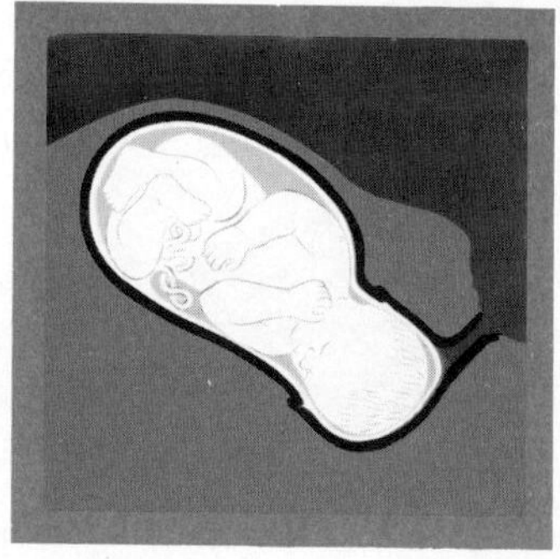

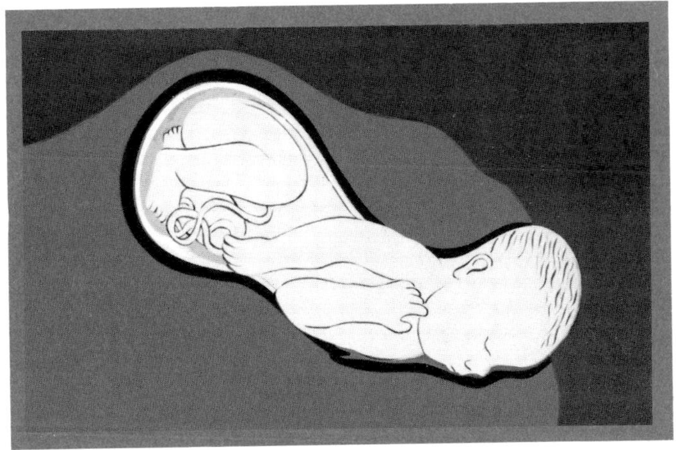

The muscle walls of the vagina expand to make room, and the baby's head moves out of the mother.

The doctor lends a helping hand as the new baby meets the outside world.

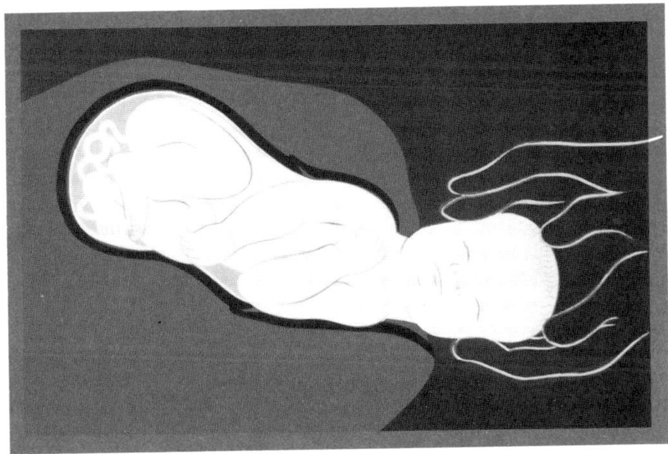

every half hour and last only a few seconds. Then, gradually, they return at shorter intervals and last longer. Just before birth, this muscular pressure occurs every minute or two.

The birth of a first child takes anywhere from five to twenty hours. Births of second and third children are usually of shorter duration. The length of time depends on the mother's build, her condition, and the size of the baby.

Whether she goes to a hospital or remains in her own bed, the mother almost always has a doctor and nurses to help with the birth. Ordinarily, the sac of fluid which has been protecting the fetus bursts during the early stages of labor. This is a signal that the newcomer is soon to appear.

Meanwhile, the organs inside the mother's body have been preparing to send the baby on its way. The set of bones in the pelvis—those which support the abdomen—stretch apart so that the baby can pass through. The cartilage which connects these bones becomes softer so that the stretching can take place.

Already, by the sixth or seventh month, the baby has settled in a head-down position. Shortly before birth, he is very active and may shift right around in the uterus. Normally the head emerges first, with the rest of the body following easily after it. The opening of the uterus widens to let the child through. Strong muscles of the uterus propel the baby out through the vagina. This passage also stretches to make room for the baby's head and body.

The infant's head is constructed in a special way so as to make birth easier. Before birth, the skull is composed of three bony parts, one on each side and one in front, with a soft spot in the middle. These bony parts are connected by flexible tissues which allow the head to be compressed, and so to travel more easily through the birth canal. They also protect the baby's brain from pressure. Several months after birth the bony parts join and the skull becomes hard.

As the mother's labor nears an end, the doctor gives her an anesthetic which prevents her from feeling the sharpest pains. Finally, as the baby's head appears, the doctor helps to ease the baby into the great outside world.

The umbilical cord—about two feet long—still connects the baby with the placenta. Since the baby will now be getting his nourishment from his mother's breast or from a bottle, the cord is no longer needed. So the doctor cuts the cord a few inches from the baby's stomach, but before doing so he ties up the cord so as to prevent it from bleeding. Since the cord has no nerves, neither the mother nor the baby can feel any pain from the operation. The baby's abdomen is then bandaged. In a few days, the stump of the cord dries up and falls off. The small round hollow left in the center of the baby's abdomen is known as the "belly button," or navel.

The placenta, and what remains of the cord, both now useless, are discharged last of all. This discharge is called the "afterbirth" because it comes out after the baby is born.

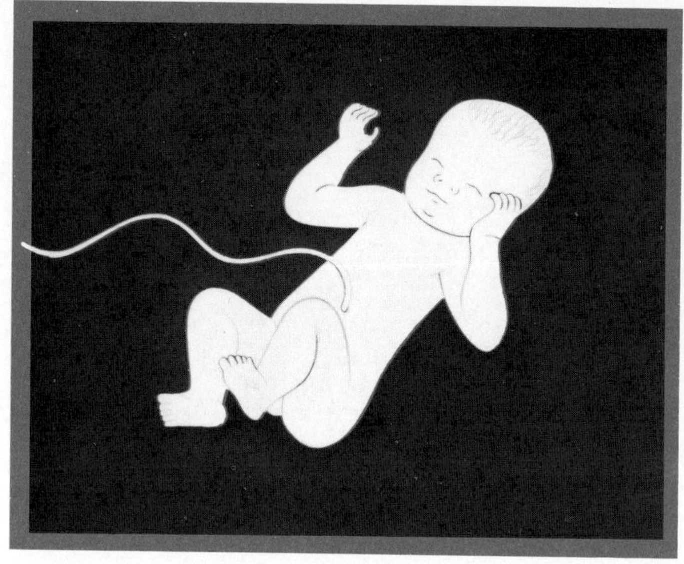

Since the baby no longer has to depend upon its mother's blood for food, the doctor ties and cuts the cord. This baby will grow to an adult and have children, thus continuing the cycle of human growth.

The baby is now a small, complete, separate human being. He opens his mouth and lets out a yowl. This first cry pleases the doctor and nurses, because it means that the baby is breathing on his own.

While the doctor takes care of the mother, his assistant attends to the baby. The infant's eyes are treated with a solution of silver nitrate, which destroys any germs that might cause an infection.

After the birth, the mother stays in bed for a few

days at the hospital, then at home. This rest gives her reproductive organs time to return to their normal condition. Although she is tired at first, the thought of being able to take care of her new son or daughter soon brings back her energy. For some weeks after the birth of her child, both mother and baby continue to see the doctor to make sure they are in perfect health.

The nine months of waiting are over at last. A father and mother have taken part in a unique creation. With it, the cycle of life begins all over again. And so human growth will continue, generation after generation, down through the centuries.

QUESTIONS AND ANSWERS

Is it all right for parents to have as many as a dozen children?

Of course it's all right. A husband and wife, however, should be sure that they can support a large family—provide all their children with food, clothing, and a decent home.

Some women, on the other hand, are not strong enough to bear a dozen children. A doctor can advise a mother as to how large a family she is physically able to have.

Wouldn't it be bad for a mother if she had a baby, then waited six years before having another one?

No. If a woman is healthy and still of childbearing age, it is quite all right for her to have another baby six years after the first one. Some children are born fifteen, even twenty years apart and are perfectly healthy.

Why are some babies kept in incubators?

Because these babies were born ahead of time—at seven or eight months, instead of nine. A seven-month baby is not quite ready for the outside world. It is too small and delicate. It needs an even temperature like that inside its mother's body, until it is strong enough to live in the outside world. An incubator protects the baby while it is weak and immature.

When babies are born, are their eyes blue?

Yes. Nearly all babies have grayish-blue eyes at birth. The eyes of most babies do not take on a definite color (such as brown or hazel) until the baby is about three months old.

Do babies have *to have milk from their mothers at first?*

No. Sometimes the mother is unable to provide enough milk for her baby. The baby is then fed especially prepared milk from a bottle.

When does the mother start menstruating again after a baby is born?

This varies, depending on the particular woman. Whether or not the mother nurses the baby also makes a difference. If she does, she may not menstruate regularly for several months or a year. The average time for menstruation to start again, in women who nurse their babies, is five months.

Mothers whose babies are not breast-fed, however, usually start menstruating between four and eight weeks after childbirth.

Does not the birth of a baby leave a big hole in the mother?

No, not at all. Remember that the baby comes out of the mother through the vagina and that both vagina and uterus are made of muscle fiber which stretches as the baby grows. After birth, certain hormones act on the uterus, which soon shrinks back to its normal size. This shrinkage is gradual. It does not happen all at once.

For how long does the flow of blood continue after the birth of a baby?

After childbirth there is a discharge from the vagina lasting from two to four weeks. This discharge is quite heavy for the first four days or so, then becomes thin and pale as it gradually decreases and finally stops.

Why are babies always spanked immediately after they are born?

They are not really spanked. Babies are given a slap, to help them start breathing. The baby's lungs expand immediately after birth, when they begin to take in air for the first time. The first cry is a signal that the baby's lungs are starting to function, that breathing has begun.

What is the "belly button" for?

After birth, the "belly button," or navel, is not used for anything. It is merely the place where the umbilical cord was once attached to the baby. While inside the mother's body, the baby obtains nourishment through this cord.

What happens to the cord when it is tied after the baby is born?

The doctor ties the cord, then cuts it. Since there are no nerves in the cord, the baby feels no pain. Several inches of cord are left fastened to the baby's abdomen. In a few days this piece of cord dries up and falls off. It leaves a little hollow place called the navel, or "belly button." What remains of the cord, which is attached to the placenta, is discharged from the mother after birth. Hence the term "afterbirth."

How long is the cord that is fastened to the baby?

The cord grows with the unborn child. At three months of age, it is about two inches long. At birth, it measures from about twenty inches to two feet. It has a twisted appearance.

What happens when a baby is born Caesarean?

Caesarean is the name for an operation performed to deliver the baby from the mother. A slit is made in the abdomen, so that the uterus can be opened and the child lifted out. Some women are not built in such a way as to give birth in the normal manner. Other factors may prevent the child from passing normally through the birth canal. In such cases, a Caesarean operation is necessary. The operation is called Caesarean because Julius Caesar is said to have been delivered in this manner.

So skilled are modern surgeons that Caesarean deliveries nowadays are usually no great risk for either mother or child.

What causes a child to be born crippled?

Babies are rarely born crippled. A crippled condition may be caused by disease, by injury at birth, or occasionally by some inherited factor.

Why are some babies born dead?

There are various reasons, some of which are: severe illness of the mother or a serious injury to her body; certain diseases; injury to the unborn child; a very difficult birth. However, the great majority of babies are born alive and healthy.

What causes freaks to be born?

Injuries, inherited factors, or gland disturbances have been known to cause freaks. It is well to remember, though, that freaks are extremely rare.

Is a baby born with its eyes closed or open?

Closed—in order to protect the eyes while the baby is moving down the birth passage. The eyes open soon after birth.

What happens when a baby comes out feet first?

Nothing happens particularly. Some babies are even born rump first. It's more normal, however, for the head to appear first.

What did mothers do when there were no hospitals?

A mother had her baby at home in her own bed. The doctor and his nurse, or some other helper, came to the home and assisted with the delivery of the child. Even today, some mothers prefer to bear their children at home. Sometimes this is necessary, since in many small towns, or out in the country, there are no hospitals.

Does having a baby hurt?

There is some pain, but it varies in different cases. When there is much pain, the doctor usually gives the mother an anesthetic. With all the new discoveries in the field of medicine, many women go through the birth process with hardly any trouble.

How does the mother know when it's time to go to the hospital?

First of all, she knows that having the child usually takes place two hundred and eighty days, or about nine months, from the start of her last menstrual period. Accordingly, she and the doctor can figure out about when to expect the baby.

Second, several hours before the baby is actually born, the mother begins to have mild labor pains, something like cramps. They occur in a rhythmic pattern, at first once every half hour or so, then at shorter intervals. Often the sac surrounding the baby bursts, and the water it contained flows out through the vagina. These signals warn the mother that she is about to have her baby.

What causes birthmarks?

Some birthmarks are spots where the pigment, or coloring substance in the skin, has thickened. This leaves a brownish patch. Red birthmarks are caused by blood vessels that have not properly developed, with the result that the blood has piled up in one place.

Somebody said that if a mother carrying a baby was scared by a snake, the birthmark would look like a snake. Is this true?

No. Such stories are merely superstitions. The baby cannot be marked in any way just because the mother is shocked or frightened. The child has its own nervous system. It has no nerve connections with the mother. It simply obtains food and other chemical substances from her blood stream.

If a woman just naturally has a big stomach, then becomes pregnant and has a baby, is she smaller afterward?

She may or may not be. It depends on her diet during pregnancy and whether or not she exercises after the baby is born. By exercising as the doctor suggests, many women acquire slimmer figures after childbirth than they had before. It varies, of course. Some women gain weight after giving birth.

If the mother smokes a lot, has it any effect on the baby?

Many doctors believe that if the mother's system is accustomed to her smoking before pregnancy, this habit will not affect the baby. It might even upset the mother to try to stop smoking all of a sudden. During pregnancy, it is important for her to be as happy and comfortable as possible.

Does anything in the mother's body change when the baby is growing?

Yes. Several things change. The most noticeable change is that the mother's abdomen sticks out. As the uterus expands to make room for the growing child, the abdominal wall, too, must stretch. The larger the baby grows, the larger the mother's abdomen becomes.

The mother's breasts also swell and become firmer. They are getting ready to provide milk for the newborn baby. Another change during pregnancy is that menstruation stops.

What is meant by a miscarriage?

This means that the baby is born before it is old enough to live. A miscarriage may take place between the first and sixth month of the baby's development inside the uterus. It sometimes occurs when the mother is ill or injured, or when the baby is not growing properly.

Doesn't the baby breathe at all inside the mother?

It doesn't breathe through the nose and take air into the lungs. The lungs do not start working until after birth, when the baby is out in the air. Inside the uterus, the child gets oxygen from its mother's blood. This oxygen travels through the placenta and the cord, along with food materials.

Does the cord connecting the baby to its mother ever break?

No. The cord is made of blood vessels and membrane, not of anything which—under normal conditions—can break.

When the egg is fertilized and attached to one side of the uterus, is the cord on that side or the other side?

The cord is on whichever side the ovum embeds itself. It connects the baby to a special organ known as the placenta, which provides nourishment. The placenta grows on the wall of the uterus. It consists mostly of blood vessels.

I read a story where people didn't have babies any more, but grew them in test tubes. Could anything like that ever happen?

Nothing like that is ever likely to happen in real life, because it is so difficult to match the conditions inside the mother's body. So far, scientists have been unable to cause a human ovum to be fertilized and to grow outside the body. It will be a long, long time—if ever—before babies are grown in test tubes.

GLOSSARY

abdomen

the section of the trunk below the chest; that part of the body containing stomach, intestines, liver, and internal sex organs of the male and the female.

adolescence

the period of life between childhood and adulthood; the time at which boys and girls start to become young men and women, as indicated by the appearance of body hair, change in voice and, in girls, menstruation.

adult

a grownup; a man or woman who has reached maturity.

birth

the act of being born; the process of leaving the woman's body and entering the outside world.

breast

the upper front part of the body, below the neck and above the waist; in women, a milk-producing gland.

Caesarean

an operation performed on a pregnant woman. The child is delivered through the walls of the abdomen.

chromosome

a tiny structure found in the cells, which carries the units that cause family resemblances. Each body cell, except the sex cells, contains forty-eight chromosomes. The sperm and the ovum contain only twenty-four chromosomes each.

circumcision

an operation performed on boys (usually when babies) in order to remove the cap or foreskin covering the tip of the penis.

cord

the ropelike structure connecting the fetus with the placenta.

cytoplasm

the jelly-like inner part of a living cell. It surrounds the nucleus.

egg cell

the female sex cell or ovum; the cell which, combined with a male cell (sperm), grows eventually into a human being.

embryo

a new life in its earliest stages; in human beings, a "baby" less than three months old, growing inside the woman's body.

endocrine

the term applied to a group of glands which secrete substances (hormones) into the blood. Examples are the pituitary, the testes, and the ovaries.

feminine

like a woman; having the qualities of a female.

fertile

capable of developing into a living thing: as an egg cell which has met a sperm cell. Also applied to men or women who are able to have children.

fertilization

the joining of a sperm cell and an egg cell, which starts the growth of a human being. Fertilization takes place inside the woman's body.

fetus

a fully developed embryo; unborn child at least three months old.

foreskin

the cap of skin covering the tip of the penis.

fraternal twins

unlike twins; those which grow from two separate fertilized eggs. They may be of the same or opposite sex.

gene

very small chemical units which make up a chromosome and which cause members of a family to resemble one another; the carriers of inherited traits.

genitals

the reproductive or sex organs on the outside of the body.

gland

an organ of the body that secretes a juice or chemical substance. The glands help in the regulation of growth and body activities. The sweat glands, for example, help to regulate temperature, while the pituitary gland affects the rate of growth.

growth

an increase in size, weight, and other features of the body; the process of developing into an adult.

heredity

the passing on of mental and physical traits from parents to children, or from one generation to another.

hormone

a chemical substance made in a gland which travels in the blood and affects the activity or growth of another gland or another part of the body.

identical twins

twins which look almost exactly alike and are of the same sex; those which develop from a single fertilized ovum.

masculine

like a man; having the qualities of a male.

mating

joining of a male and female in the reproductive act; the placing of the penis into the vagina, so that sperm cells can be deposited.

maturity

adulthood; complete development or full growth.

menopause

the period in middle-aged women when menstruation ceases, usually in the middle forties.

menstruation

the normal flow of blood from the uterus through the vagina, occurring once a month for most women.

miscarriage

the birth of a child before it is old enough to live, between the first and sixth month of its development in the uterus.

nucleus

the center spot or core of a cell; the round part which contains chromosomes.

ovary

one of the two female organs which produce egg cells or ova and also certain hormones.

ovulation

the process of discharging eggs from the ovary; the production of ripe ova.

ovum

the female sex cell; the egg cell. If joined by a sperm cell, the egg develops into an embryo.

penis

the male sex organ through which sperm cells leave the body; also used to discharge urine.

pigment

substance in the cells of the body which gives color to the hair, skin, and eyes.

pituitary

a small gland located in the brain, which secretes the growth hormone and helps to regulate body activities.

placenta

a flat, spongy structure which grows inside the uterus during pregnancy to aid in the exchange of food and body wastes between the mother and the unborn child.

pore

tiny opening in the skin through which perspiration passes.

pregnant

expecting a child; a woman with an embryo or fetus in her uterus.

puberty

the time when a person reaches sexual maturity, or is able to reproduce; at about the age of twelve for most girls and fourteen for most boys. Puberty is marked by growth of pubic hair and hair under the arms, change in voice, and—in girls—development of the breasts.

pubic region
: the area at the extreme lower end of the abdomen, where the sex organs are found.

scrotum
: the sac of skin containing the male testes.

semen
: the whitish fluid containing sperm cells which is produced in the male reproductive organs or testes.

seminal emission
: the discharge of semen from the penis during sleep, often accompanied by a dream, a "wet dream."

sex cell
: the sperm in the male, or the ovum in the female. These cells combine to start a new life.

sexual intercourse
: mating; the uniting of the penis and the vagina.

sperm
: the male sex cell which, by joining with an ovum, starts a new life.

sterile
: a man or woman unable to become a father or mother.

testis

one of the two male organs which produce sperm cells and hormones. The testes (plural) are located in the scrotum.

umbilical cord

the ropelike structure connecting the fetus with the placenta.

uterus

the organ inside the lower part of the woman's abdomen where a baby develops before birth; the womb.

vagina

the passage from the uterus to the outside through which babies are born; the birth canal; the place where sperm cells are deposited during mating.

vulva

the outside or visible parts of the female sex organs; the entrance to the vagina.

womb

the uterus; the organ which houses the unborn baby inside the woman.